Paleobotany

Paleobotany

Daniel Dela Torre

Delve
PUBLISHING

www.delvepublishing.com

Paleobotany

Daniel Dela Torre

Delve Publishing
2010 Winston Park Drive,
2nd Floor
Oakville, ON L6H 5R7
Canada
www.delvepublishing.com
Tel: 001-289-291-7705
001-905-616-2116
Fax: 001-289-291-7601
Email: orders@arclereducation.com

ISBN: 978-1-77361-298-0 (Hardcover)

ABOUT THE AUTHOR

Daniel completed his MSc in Environmental Science from the University of the Philippines Diliman in 2016. His primary interests are ecological monitoring using remote sensing and the effects of climate variability on vegetation. He is engaged in the University of the Philippines as a Senior Science Research Specialist working on LiDAR mapping of agricultural resources.

TABLE OF CONTENTS

LIST OF FIGURES

PREFACE

There are important and very interesting branches of science, of which little is known to the public. This is what happened with paleobotany. From time to time, there are popular books and articles about extinct animals, but fossil plants are unlucky. Books that tell what paleobotany is, who are paleobotanists, what they work for and why, can be counted on fingers because the number of them is so low. Meanwhile, the history of the plant world is no less instructive than the history of the animal world. Paleobotany has a huge store of information that is of great importance for solving both general biological problems and purely practical questions of geology. Without active participation of paleobotanists, it was not possible, for example, to master all the coal basins. However, the study of fossil plants is not only important, but also very interesting, even fascinating.

Written at a high scientific level and at the same time quite popular, this book will undoubtedly be interesting and useful to anyone who is interested in the distant past of our planet. However, it is not only that, but this century is the age of atomic physics, space flights and molecular biology where many seem to be focused on the future. Paleontology (and even more so paleobotany) is becoming more and more the focus of some eccentrics and is farther and farther away (and maybe - and Just lagging behind) "real" science, the full movement of scientific thought, the burning and fights of brilliant hypotheses and ideas. This book is a vivid answer to this deeply erroneous view.

This book covers a wide range of issues. It tells about the most ancient plants and modern subtle methods of studying them. In addition, this is not a popular translation of a paleobotany textbook. In this book it is shown that the vegetable world of the past is important, but also how facts were obtained, and important generalizations were born. The reader will become, as it were, a witness of many years of discussions in which paleobotanical data played an important role. The author touches upon the most general theoretical problems, having expressed many original thoughts on them.

There is no doubt that not only paleontologists, paleobotanists and geologists, but also the widest circle of readers, from physicists to lyricists, will read the book with interest and benefit.

Author

1 THE FIRST SETTLERS

GOETHE

Since the ideas of evolution began to penetrate paleobotanical study, the earliest pages of the history of the plant world have not left researchers alone. That feature of human thinking, which is called "spontaneous dialectics," made paleobotanists look more closely at the most ancient "documents" to understand how it all began. However, the oldest remains of plants, for all their scientific value, are still the most mysterious. While scientists already know a lot about them, the main question remains unanswered. We cannot even say approximately how and what plants first came to the land, whether the plant kingdom had its own Adam and Eve, or whether it had many basic roots. From the question "why?" Now it is better to give up altogether. Answering it, we cannot yet escape beyond the limits of logically possible, but not observed, options.

With the study of the first land settlers, paleobotanists did not hurry. Looking at scraps for a long time did not wake the imagination, which without fine techniques of research had nowhere to go. More often, it happened like this: find an imprint, describe it under a new Latin name and return to the usual collections. This story is familiar to archaeologists. How many it took years that the remains of prehistoric people should be given as much attention as to the monuments of ancient Rome!

A turning point in relation to the first land plants came only in the early 20s of the 20th century. Everything settled the case. In the place of Rainey in Scotland, geologist Mackey discovered a piece of rock with the remains of plants of amazing quality in a stone fence. The geologist, who knew well which rocks are found in the area of Raini, quickly managed to find the place from where the stone was born. A lar: R. Kidston and V. Lang. Within five years (1917-1921), descriptions of the plants found in Raini were published, and every paleobotanist now knows about them. From these plants, textbooks, reference books and popular books were published (Gilbert, & Willerslev, 2007). We, too, will not miss the chance to talk about them, but not to maintain tradition, but for the simple reason that Scottish finds are still the most revealing and irreproachably studied.

PETRIFIED SWAMP IN RAINI

In the first half of the Devonian period (almost 400 million years ago), when reservoirs were laid in the Urals for the oil fields of the Second Baku, Rainy had a small swamp with numerous but monotonous plants (Fig. 1). Kidston and Lang counted three kinds (not counting one alga) here. Two of them, rhinia (Figure 2) and the mountaineophiton, had low, not more than a meter, bare stems with creeping rhizome at the ends of the stems - small pieces with microscopic sporangia (sporangia), which were sown and gave rise to a new generation of bare stalks. The third genus (asteroxylon) had stems with small siliceous leaves. The internal structure of all these plants was just as simple as the appearance.

Figure 1: European landscape in the early Devonian era.

Psylophyton was found in the Devonian rocks of North America and described by Dawson as far back as 1859, but only 60 years later, they remembered this discovery, which was hardly noticed at the time (Gilbert, & Willerslev, 2007).

Figure 2: Rinia from the Devonian deposits of Scotland: a - reconstruction of the external appearance, at the ends of the branches sits sporangia

Figure 3: Rinia from Devonian deposits of Scotland: b - incision of sporangia, filled with spores.

The publication of Scottish material has become a sensation for paleobotanists and for all who are interested in the history of plants. It seemed that the veil with one of the main mysteries was sleeping. In the theoretical works, the genealogical trees of the vegetable kingdom were glimpsed, in the roots of which the psilophyton and the trio from Raini inevitably stood (Doyle, 2000). The hobby was so significant that it somehow eluded the age of the Scottish finds. After all, they are not the oldest. In earlier deposits, too, there are plants, and much more complexly arranged ("advanced"). Yet the skeptical mood that accompanies every sensation has gradually surfaced. In 1954, Belgian Suzanne Leclerc, an authoritative expert on the most ancient plants, came up with an article very symptomatically titled: "Are psilophytes the starting point or the endpoint?" At first glance, it is a ridiculous question. What end point can we talk about when we have primitive ancient stems with an equally primitive inner structure? Continuing the analogy with the archeology of prehistoric times, the question of Leclerc can be pictured like this: "Were the ancient stone uncouth choppers and hammers the initial or final point in the development of human instruments?" However, this analogy is already illegitimate. Skepticism in the views of Leclerc and her associates was not accidental, although at first it was taken seriously by a few (Mclachlan, 2007).

WHY DID SUSANNA LECLERC DOUBT?

Among paleobotanists (as, perhaps, in general naturalists), who left a significant mark in their science, there are two varieties. Some studied many collections from different parts of the globe, left a lot of work, directories, textbooks, organized congresses, and headed schools. Whatever plants the paleobotanik was engaged in, it will not surpass the works of such predecessors, the sphere of their influence was so broad. The second variety of paleobotanists is those who, throughout their life, worked on a relatively narrow topic and painted a picture with exhaustive detail. They are not always widely known, but they laid the foundation for paleobotany.

Suzanne Leclerc belongs to the second type of researcher. For some time she studied the anatomical structure of the Carboniferous plants in coal kidneys (we will talk about them in a seperate chapter), and then switched to the Devonian plants of Belgium, which had been engaged in a good three decades already. Articles and monographs written by Leclerc provide an indispensable guide for all experts on ancient plants. She made the most important discoveries without particularly complicated techniques. The main tools in her hands were a strong steel needle and a small hammer.

Figure 4: Micro-fragments of the sporangia of the foreground.

Remains of fossil plants - not always flat, engraved on the rock pattern of veins.

Sometimes a branch with sporangia and small knots was gradually tightened with silt and retained its original volume. On the cleft of the rock a slice is seen in one plane, and what is done inside the sample is unknown. Armed with a needle, a camera and a truly feminine patience, S. Leclerc went millimeter by millimeter to reveal the breed, looking for the branch or difficult branching leg sporangia. She carefully sketched and photographed all the stages of her microranoscopes. Gradually the volume model of the plant was revealed in all detail (Figures 3 and 4). Although Leclerc studied the peers of the Scottish plants and even their predecessors, now and then much more developed plants, far from primitive forms, arose (Doyle, 2000). She and other paleobotanists studied Devonian plants from different places. Some plants are depicted in Fig. 3 and 4. The stems are whimsically branching, the sporangia are collected into neat spikelets, and the inner part of the trunk is filled with complex tissues. Sometimes more complex plants are credited as paleobotanists to psilophytes, but this is done more out of respect for their age than on the basis of botanical arguments. However, there was a very peculiar situation. The most ancient plants, by their complexity, were superior to the Scottish ones, or, at best, stood on the same level with them.

WHAT WERE THE FIRST SETTLERS?

Careful researchers, who are not inclined to quickly give up their current views, are now looking at plants from Rainey as long-surviving prototypes. They, researchers say, have found a suitable shelter in which they hid themselves from competitors better fitted in other places. Accordingly, psilophytes still seem to be the inevitable stage, which the plants somehow had to pass, on their journey migrating to land (Franzke, et al., 2011).

Many paleobotanists, perhaps even the majority, adhere to the point of view that plants of the Rhinia type (but not necessarily rhinia from Raine) are the starting point in the development of all terrestrial plants. In the course of evolution, the shoots became more and more branching, the main trunk gradually emerged, which acquired a more complex and different internal structure than the lateral branches. The lateral branches were flattened, fused and turned into leaves. In various directions, sporangia also moved. Therefore, we come to more and more complex descendants, now populating the continents. In addition, the swamp in Raini may have simply turned out to be a reserve for the grandparents (Doyle, 2000).

Supporters of another point of view (Leclerc belongs to this group) cite an example of a very remarkable tropical plant psilot (Psilotum, Fig. 5). It is a liana with thin branching shoots, small leaves, its internal structure is a little more complicated than that of the Rhine, but one detail breaks the harmony of simplicity. Sporangia psilota is collected by three, fused with walls and placed in the sinus of the two-horny process. Fused sporangia, moreover, sit in the bosom - this is too much and it is difficult for a plant that poses as a relative of psilophytes. According to many botanists, the ancestors of psilot were significantly more complex, but went a long way of simplification.

Figure 5: Modern psilot, which looks very similar to psilophytes.

With such amateurs, it is very often necessary to discard insignia and dress up in something simpler than a biologist. Sometimes the external simplicity of the structure is a higher level of organization behind strict, laconic forms, a powerful and perfect mechanism can hide. However, sometimes-simple living conditions make old improvements unnecessary, and the body quickly loses them (Franzke, et al., 2011). This second case, some paleobotanists see in the Rhine - as well as in psilot. However, such skeptics have appeared for a long time. Shortly after the publication of the last part of Kidston and Lang's work, Englishman DG Scott, one of the most thoughtful paleobotanists, wrote in 1924 that simplicity of the rhenium is hardly primary, since algae (and terrestrial plants could only originate from them). There are forms with already well differentiated "leaves." This is also a serious argument against the initial point. In addition, the most serious argument (Thomas et al., 2008).

Botanists have long noticed three large, almost immiscible groups of terrestrial plants, more complexly organized than mushrooms, lichens and mosses. In the very first approximation, these three groups can be defined as follows: plauniform, various horsetails (and their relatives) and all the rest. Plaque-shaped leaves are usually small (exceptions are rare), they sit spirally, and there is not much wood in the trunks. In horsetails with relatives, the leaves are also small, but they sit with whorls; there is little wood in the trunks, but here the trunk itself and the branches consist of separate segments, hence the name of the group "arthropods." The third group includes ferns and all seed plants. Often they are called megafilic, that is, large-leaved, but conditional, since their leaves can be very small. Without resorting to modern botanical terminology, it is difficult to show the differences between these three groups, and therefore their differences seem slightly artificial to the uninitiated reader (Gilbert, & Willerslev, 2007).

Some paleobotanists are inclined to think that only the third group of plants originated from the rhenium-like ancestors, and the first two (placenta and arthropods) appeared independently. On the side of this point of view, more and more facts are now gathering. For a long time no one doubted the antiquity of the first group, it meets with the lowest Devonian; this is for sure. Alternatively, maybe it existed even at the beginning of the Paleozoic Era. However, with the origin of the craniocerebals there is still much to be understood. It was believed that their ancestors were psilophytes and two intermediate genuses have even been found. The latter were given great importance, they were told about to students. Recently, the same Leclerc, together with the

German paleobotanist I. Schweitzer, debunked one race. It turned out to be a fern. Then the Americans R. Bonamo and H. Banks questioned the second kind. The connection of arthropods with psilophytes hung on a thread, and the point of view on the independence of all three groups was proportionally strengthened (Wilf, et al., 2003).

What is the result? The ancestral position of the Scottish plants is in doubt. There emerged some independent groups, three "Ivan, who does not remember the kinship." There was a situation very characteristic of modern paleobotany, and perhaps also of other areas of natural science. It was well said by the American G. Andrews: "Perhaps the greatest contribution that paleobotany made was not filling the gaps in our knowledge of the evolution of the vegetable kingdom, but showing us how many gaps exist," (Gilbert, & Willerslev, 2007).

ANOTHER GREAT MYSTERY OF NATURE

Therefore, we do not know how terrestrial plants originated, we do not know their ancestors, but maybe we can at least assume when they came to the dry land? Unfortunately, this question remains unanswered. In the rocks of terrestrial or coastal-marine origin, plant remains are not uncommon. Descending from the youngest rocks "down the slit" (as geologists usually say), we find whole "cemeteries" of leaves, seeds and stems. If there are no such remnants immediately visible, it does not mean anything: very often, many microscopic shells of fossil spores and pollen are preserved in the rock. They are extracted by special methods (this can be read in SA Safarova's popular pamphlet "With a microscope into the depths of millennia," 1964, "Nauka" publishing house). This continues until the Devonian deposits. Devonian plants known to us are abundant and diverse. The fact that they are not clear is another matter. However, here we enter into the deposits of the very beginning of the Devonian, and immediately begins the dead kingdom. Here and in the Silurian strata below, the remnants of plants are rare, unattractive, and for the most part unusual even for connoisseurs of the Devonian. Disputes still occur, and then they disappear. In Ordovician and older sediments, treated with acids and alkalis, it is possible to find some microscopic shells, but there are more discussions about them than reliable data that can be interpreted somehow. At first, they were mistaken in the peculiar controversy of terrestrial plants, and the imagination began to paint green land almost in Precambrian times. The images of these shells showed a characteristic three-beam gap, through which the spores germinate. Then doubts arose in the correctness of the images, the three-beam gaps began to disappear from them. Now most paleobotanists believe that all these most ancient shells (perhaps with rare exception) belong to algae. Such incomprehensible "documents," which have not yet been subjected to detailed study, we meet in the Ordovician, Cambrian and older deposits. The situation is completely incomprehensible. Where does the diversity of the Devonian flora come from? Kindly only a variety! Devonian plants already have a complex internal structure, which for the subsequent history of the plant world, has changed only along the line of secondary reorganizations. In the post-Devonian period, the general appearance of plants and the organs of their reproduction changed the most. However, here the Devonian plants have already reached a great deal.

In recent years, paleobotanists have been stubbornly searching for new "documents" of the pre-Devonian plant history, but the successes so far are negligible. At the same time, more and more highly organized groups are found in the Devonian. So to the riddle: "From which did the first terrestrial plants originate?" Add to this a new, no less complex, puzzle, fully asked only recently: "Why do plants appear suddenly and immediately highly developed in the Devonian deposits?" (Thomas et al., 2008). Unexpected answers to the secrets of nature can be sensational. However, the newly emerging question is sometimes even more sensational, at least for a specialist. Paleobotany faced, therefore, a problem that is worth the other great mysteries of nature. Will there be an answer to it? It will probably not happen soon. We have to rely on unexpected finds, or on a long analysis of all the circumstances of that incomprehensible yet crucial moment. All will have to take into account who will take up the issue: the features of the accumulation of precipitation of the time, and the reports of experts on fossil disputes, and much, much more.

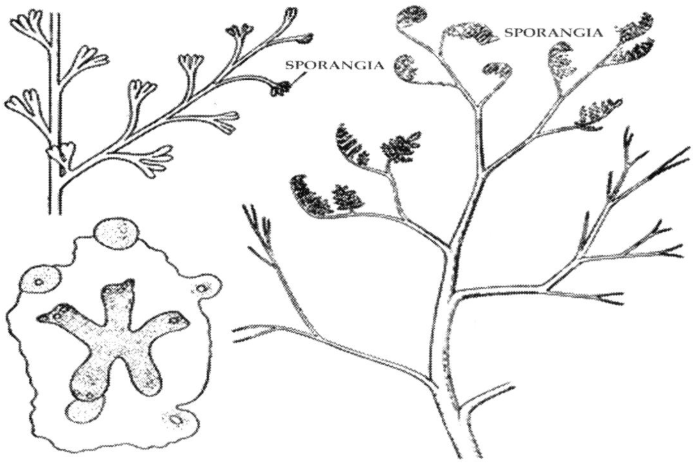

Figure 6: Devonian prapaportiik (Rhacophyton) with a very complex branching of the stems.

REFERENCES

1. Gilbert, M. T. P., & Willerslev, E. (2007). PALEOBOTANY/ Ancient Plant DNA. *Encyclopedia of Quaternary Science*, (1916), 1574–1581. https://doi.org/10.1016/B0-44-452747-8/00213-1
2. Thomas N. Taylor, E. L. T. and M. K. (2008). Paleobotany The Biology and Evolution of Fossil Plants. *Paleobotany (Second Edition) The Biology and Evolution of Fossil Plants*, 0–1. https://doi.org/10.1016/B978-0-12-373972-8.00029-2
3. Wilf, P., Cúnego, N. R., Johnson, K. R., Hicks, J. F., Wing, S. L., & Obradovich, J. D. (2003). High plant diversity in Eocene South America: evidence from Patagonia. *Science*, *300*(5616), 122–125. https://doi.org/10.1126/science.1080475
4. Franzke, A., Lysak, M. A., Al-Shehbaz, I. A., Koch, M. A., & Mummenhoff, K. (2011). Cabbage family affairs: The evolutionary history of Brassicaceae. *Trends in Plant Science*. https://doi.org/10.1016/j.tplants.2010.11.005

5. Mclachlan, J. S. (2007). PALEOBOTANY/Paleophytogeography Paleophytogeography. *Encyclopedia of Quaternary Science*, 1594–1598. https://doi.org/10.1016/B0-44-452747-8/00188-5
6. Doyle, J. A. (2000). Paleobotany, relationships, and geographic history of Winteraceae. *Annals of the Missouri Botanical Garden*, *87*(3), 303–316. https://doi.org/10.2307/2666190

2 300 MILLION YEARS OF CANNED FOOD

Chemical tables, microscopes, incomprehensible devices, and white coats of employees... Unfortunately, the situation in paleobotanical laboratories is somewhat more prosaic. The room is lined with cabinets with collections and books, tables and boxes with stones, heaps of photos and papers, in the corners, parcel boxes with samples sent by geologists from various places from the corner draw unpleasant smells (a new Method of processing prints of plants has been mastered). The science of fossil plants has always been a poor relative living in an uncomfortable corner. However, there were always enthusiasts, thanks to whom we now know a lot about the vegetable world of the past (Gayangos & Brett, 2002).

The prints of plants that come to the pages of scientific and popular publications are few lucky ones from hundreds of thousands of their colleagues. Most of the material passing through the hands of paleobotany is much more unimpressive. About archeology it is sometimes said that it studies, mainly, the trash heap of past civilizations, that this is "the science of broken pots." Paleobotany also rarely gets "gold bracelets and pearl necklaces." It is not a simple matter to see a tree behind scraps of leaves and pieces of trunks. There is little patience and botanical knowledge (without them, of course, it is better not to take up work). We also need modern technical methods. The latter will be discussed in this chapter. It is not necessary to think that all the methods to be discussed are an indispensable companion for any study. As an investigator does not necessarily use the whole arsenal of criminalistics, so in paleobotany it is not always necessary to apply all the achievements of its science in everyday practice (Gayangos & Brett, 2002).

WHEN A MINER SWEARS, AND PALEOBOTANY REJOICES

The coal layer sometimes surprises miners. In the very thick of the coal, from nowhere, there are heavy black cobblestones of plain stone (Fig. 6). You cannot burn them, there is no use and there is a lot of trouble. Sometimes they are found in such quantities that there is almost no room for coal itself. The size of these cobblestones ranges from very small to very large. There are cases when, for several meters, the coal seam is completely replaced by one large joint of a strong stone weighing several hundred tons. Miners swear, but for paleobotany, such stones are a real treasure. The English call them "coal balls", the Germans - "peat dolomites," and Russian literature has become more familiar with the name "coal kidneys." The miners of the Kuznetsk Basin call them "scammers." The shape of coal kidneys are very diverse. They can be accurate balls, and ridiculous flat cakes with outgrowths in different directions. Collecting coal kidneys, a paleobotanik does not know what he has found. All will decide the first cuts. The stone may be empty, but more often an amazing sight opens up.

Figure 7: Appearance of a small coal kidney.

Coal buds are unique plant cemeteries, a real "petrified vinaigrette" (Figure 7). Even before the accumulation of plant residues turned into a coal seam, the salt solutions preserved the stems, leaves, cones and seeds of plants with their entire cellular structure. Then the time for them seemed to stop (Dilcher, 2001).

The first paleobotanists, who took up the study of coal kidneys, prepared transparent sections of them - polished sections. Petrographs do this when they want to study the structure and mineral composition of the rock. The stone is split or sawn. The resulting plane is leveled on a grinding wheel and glued with a special resin on the glass. Then the stone is sanded from the backside, until from it on the glass there is not only a thin transparent layer left. The thickness of the section is several hundredths of a millimeter. Of the coal buds, the paleobotanists first cut out a series of parallel plates, and each made a thin section. Therefore, it was possible to trace the structure of the buried trunk, leaf or bump in several places (Dilcher, 1991).

The coal kidneys opened a whole world to paleobotanists, a completely new one for them. Instead of prints and pressure-packed rocks of blackened "mummies," carefully prepared "canned food." Fossil plants of coal kidneys look almost alive in the slices. It is not necessary to explain how the paleobotanists seized the opportunity. Some have devoted their lives to studying these kidneys of coal.

HOWEVER, IS IT POSSIBLE WITHOUT THIN SECTIONS?

The advantages of coal kidneys in front of other cemeteries of fossil plants are obvious. However, there are shortcomings in them. In the kidney, if it is split, at first nothing is visible. Matte dark-brown, almost black, earthy mass, only in some places weak contours are outlined. It is necessary to make a thin section. However, it is nice to grind a small piece the size of a nickle or a matchbox. On the other hand, try to make a thin section of a hefty stone the size of a head! It is necessary to cut out a thin plate, to make it perfectly even, so that it will fit evenly to the glass. It is even more difficult to grind a large glued on the glass plate to a uniform transparent layer. Specialists-grinders, who are able to do this kind of jewelry work, are only a few. To understand the structure of the plant, one cut is not enough, we need a series of successive sections in different directions. However, that is not all. It is not enough to study a plant, it is necessary to study the biological species, which means that it is necessary to search and cut several specimens. In addition, how insulting to spend a lot of time on one sample, having several boxes of stones, each of which is a cat in a bag and can be hiding something much more interesting. It was necessary to find some way to do without complicated and expensive polishing (Carcaillet, 2007).

This method was invented by Scot John Walton. His thought worked in this direction. The plants in the coal kidney are impregnated and cemented with lime cement, which is easily dissolved with acid. If the ground surface is poured with hydrochloric acid, the cell walls compiled by the organic material will appear outward. The acid does not hurt them. Now the contours that have emerged from the rock must be cut without breaking. You cannot do this with a knife. Tissue cells, although well preserved, without the support of the cement that has held them, will at first touch crumble in the dust (because they are almost 300 million years old). Walton tried to pour etched surface with a special solution, similar to nail polish. After it dried, a thin and transparent film remained on the capillary, into which all the cell walls that have emerged from the etching were soldered. The film was elastic, durable, transparent and easily removable from the rock.

Then they guessed to do something even easier: the etched surface was poured with a solvent and, not allowing it to dry, rolled a thin sheet of film from above. Its lower layer immediately dissolved and seized the cell walls. Such films turned out to be very smooth, and they do not cost anything. This new method for studying coal kidneys (it is applicable to other types of rocks with mineralized residues) has had a striking effect. At the same time, several samples are being processed for processing. Some are poisoned, others dry after pickling, and others are already covered with films. Thousands of impressions began to appear from such conveyors. If earlier the paleobotanik could not make several slices at a distance, say a millimeter from each other, now it became possible to cut at any interval. From a millimeter of a coal kidney, it is possible to prepare dozens of impressions. So sometimes they do. Try differently to study in detail the structure of a handful of sporangia, each of which is 1-2 mm in size, the legs are even thinner, and, moreover, they are intricately branched (Carcaillet, 2007).

PALEOBOTANY-CONSTRUCTOR

In one of the numbers of the Czech magazine "Photography" was placed a photo-book. It shows an increased cut of cabbage head. It was not immediately clear alternation of white and dark stripes. Mystery, probably the quickest guess was the homemakers, to whom the sight of the head in the section is not a curiosity. It is not difficult to recognize a watermelon or an apple in the cut. However, when you see a section of an unknown plant before your eyes, it is not easy to understand at once, what is what. An engineer or a skilled worker can easily draw a general view of the three sections of the detail. In addition, try to understand a complex machine of unknown design, if you are offered several hundred slices. To do this, of course, it is possible, but how much effort should be put to trace from one cut to the next, what is connected to, and what is the shape and structure of each part. This is the work of the paleobotanists studying the contents of the coal kidneys (Fig. 8).

Figure 8: Reconstruction of plants from coal kidneys.

For the reconstruction of difficultly branching stems, an interesting method is sometimes used. Increased contours of successive slices are applied to sheets of transparent plastic or wax, which are then superimposed and glued together. In a transparent pile o, you can clearly see the contours passing inside, you can cut out the enlarged model of the plant. Approximately this way it was possible to understand the structure of the ancient fern Stauropteris from the Lower Carboniferous sediments of Scotland. Remains of stavropters were, however, found not in coal buds, but in calcareous sandstone (Manchester, et al., 2009).

In coal, kidney paleobotanists hoped to find answers to many questions. Indeed, they were able to obtain immense amounts of information about the plants of the coal-and-coal forests. However, often it happened that the answers received resembled new, even more complex issues. Here is a typical example.

Cutting coal kidneys found in one of the Illinois deposits, the American, Sergius Mamay, discovered a large, cigarette box -sized cluster of sporangia of an ancient fern. A single sporangium had a millimeter in diameter, four millimeters in length and looked like a tiny banana (Figure 8, a). The wall of sporangia consisted of several layers of cells, which is rare in modern ferns, and in Carboniferous time was not a wonder. However, the Illinois fern (it was called biscaliteka) was interesting. Usually sporangia have a wall consisting of identical cells. Only a small group of cells is thickened. This device is for opening a mature sporangium. In biscitants, thickened cells are located in two broad bands, and in the rest of the wall, thin cells alternate with groups of small-thickened cells. In a word, a very complex structure. Perhaps, in the whole vegetable kingdom there are no sporangia with such a clever wall. However, even this was not the most amazing thing.

Usually in ferns, spores in sporangia mature, are sown and, if carried with good soil, sprout. Tiny plants (sprouts) are produced on which male and female genital organs develop (Dal et al., 2009). A new fern grows from a fertilized egg. In bizkalitiki, apparently, it was not so. In petrified sporangia, Mamay found already sprouted spores, and in some cases, it came to an outgrowth of 10 cells. What happened to these sprouts then, as a new fern grew, we do not know. In the piece of the coal kidney, only one stage of the complex life cycle has frozen, and the paleobotanists have received a task that, if understood, costs more than the notorious problem of the "snowman."

SIMULATION PLANTS

In biology there is such a thing as the simulating forms. This is not the first time that the everyday and scientific meanings of the word do not coincide. "Simulating" in biology does not mean, "portraying an ailment." Simulating forms are simply outwardly similar animals or plants that are not really relatives. For such similarities, many other names have been introduced: homoplasia, convergence, parallelism, homeomorphy, mimicry, etc. However, this is already a terminological jungle, in which it is better not to climb.

Probably, many beginning mushroom pickers flung out a vale, which looks like a white fungus from a distance. If such a mushroom did not get into the pot, this ends up a conflict with the simulating form, it is enough to look under the mushroom's head, and the simulation is exposed. However, in paleobotany is not always so easy to look under the bonnet. The plant almost never falls into his hands in its entirety. Usually it is only stems, or only leaves, seeds, cones, roots, etc., and they can be very similar in different plants (remember the nettles and the stinging nettles, which belong to different families). However, sometimes it is the other way round: at first glance, you do not recognize close relatives for anything. An inhabitant of the middle band in the sheet in Fig. 9, b does not recognize the oak, yet it is an oak, which also has acorns. For an expert in modern plants, such anomalies are not a problem. He sees the plant as a whole, "with acorns." For a paleobotanik, studying prints, among which such oaks can be caught, also sees a catch if he knows modern plants. However, as soon as it starts to descend into the depths of geological epochs, it cannot do with one knowledge of modern plants. Paleobotany must look for some important marks on the fossil leaves, distinguishing or, on the contrary, uniting plants in natural systematic categories (Bradshaw, 2007).

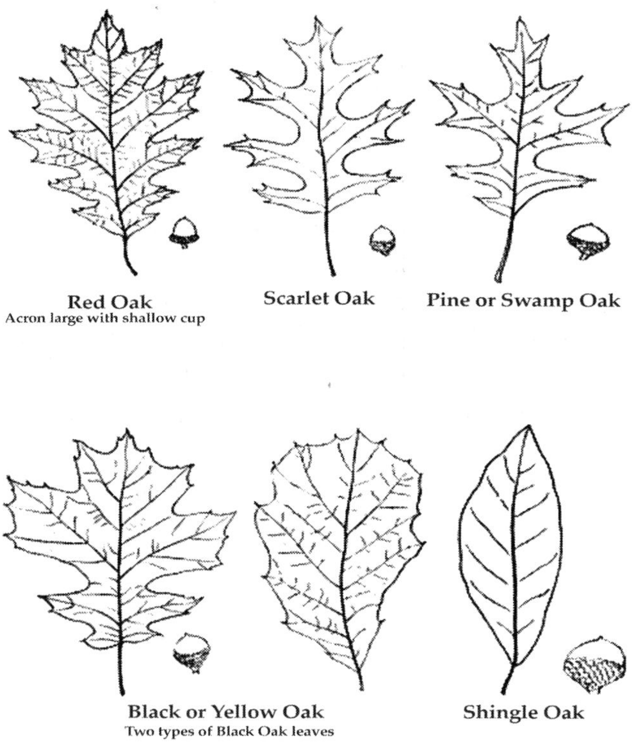

Figure 9: An example of a different form of leaves belonging to the same genus.

It is not accidentally inserted the word "important." Similarity and difference are extensible concepts. All leaves are somewhat similar, and at the same time, there are no two identical. Speaking here about the differences in leaves, we mean their belonging to different systematic categories. Two sheets of the same species have intraspecies differences, two sheets of different species - specific differences. Next, come the distinctions of generic, family (from the systematic group "family") and so on to the type. All these categories - species, genus, family, etc. - are established in botany on the plant as a whole, everything is taken into account, but the main attention is to the reproductive organs (Bradshaw, 2007). They are least susceptible to accidental changes and therefore are most reliable for taxonomy. This is where the main troubles of paleobotany begin, to which plants come in "disassembled" form, with broken and completely lost parts. How to assess the observed differences and similarities, what to give preference?

CHEMISTRY AND MICROSCOPE ASSISTANTS

Once the appearance is deceptive, since two similar sheets can belong to different species, genera, families, orders and even classes; and, conversely, once two leaves that have little in common can belong to the same species, genus, etc., then we will leave the appearance aside and look inward, into the detail, into the microstructure. Perhaps, about a hundred years ago German paleobotanist Johann Bornemann reasoned this way, introducing into his science a new method, which was later called "epidermal," "cuticular," or "epidermal-cuticular."

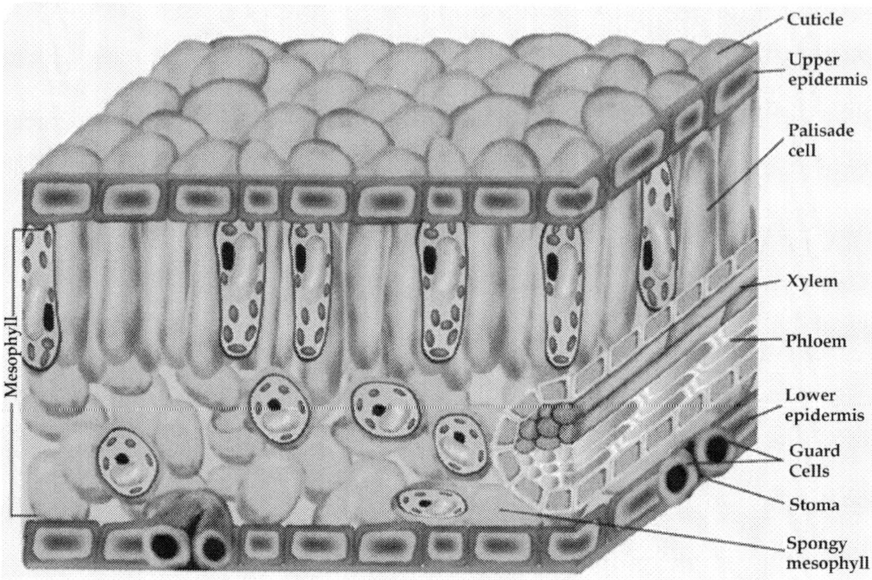

Figure 10: Cuticle (epidermis) of various fossil plants.

Even in the first half of the last century, botanists were aware that the surface skin (usually called the epidermis) of leaves of different plants was not the same (Figure 10). Hairs, tubercles, the shape and size of cells, the fine structure of the walls separating the cells - here are a few taken at random features that distinguish the epidermis of different plants. However, the main characteristic property of the epidermis is the stomata (Figure 11). With the exception of fungi, lichens, algae and mosses, stomata are found in all plants, and their diversity is inexhaustible (Crane, et al., 2004). Through the orifices of the stomata, gas exchange and evaporation of excess water occurs. The stomata itself is not just a hole, but a miniature relay, sensitive to the humidity of the air and the water content in the body of the plant. It became too dry - all the stomata are closed, only the attendants allow the plant to breathe, but it has rained, and the cells that close the holes diverge.

In most terrestrial plants, the epidermis is covered from above with a thin film of high-polymer organic matter of the cutin, which does not allow gas exchange to go anywhere, besides stomata. This film - cuticle - neatly repeats the outlines of all the cells of the epidermis, hairs, stomata and itself is sometimes decorated with a thin ornament or is provided with pores. Of course, it is good when you can study both the epidermal cells and the cuticle, but on fossil leaves, it is rarely possible to see both. Fortunately, the cuticle is a good copyist, you can see almost everything you need on it. It is only necessary to remove it from the fossil sheet. In addition, here, finally, nature has come to meet the researcher - a rare case, which the paleobotany used (Gayangos & Brett, 2002).

Unfortunately, in their articles in the form of a dry report, scientists rarely write how they came to the discovery. For this reason, it is now difficult to reconstruct the idea of separating the cuticle from fossil leaves. Perhaps it was so. Nature itself did it first. The cuticle-forming cutin is a very persistent substance. Almost all known acids and alkalis are afraid of it, time also practically does not work on it. The main enemy of the kutine is high temperature. It may be fatal and prolonged exposure can cause weathering. However, before getting to the cutin, weathering usually destroys the charred interior of the leaf. As all this happens in nature, we do not know for sure, but only in the rocks, there are leaves or their scraps, from which only the cuticle shell remains. Sometimes veins remain in such a flat bag (the cuticle covers the sheet from both sides). Such remnants of nature treated by nature fell into the hands of I. Bornemann. He examined the film under a microscope, found a stomata, cells with tubercles, carefully described everything and accompanied the work with excellent color patterns. Bornemann realized that all the details he had noticed could be used to determine fossil leaves, but before that, he had tested modern plants on the same subject. He also included in his monograph drawings of the microstructure of some of them.

The first experiment turned out to be successful. Now it was necessary to do with other leaves the same as nature did with the specimens of Bornemann. Here, chemistry came to the rescue. It was necessary to dissolve the charred part of the leaf and not damage the cuticle at the same time. By that time, i.e., by the second half of the last century, chemists had already learned how to dissolve coal. It was subjected to deep oxidation, and then the resulting compounds were converted to alkaline solution. Let us now mentally trace the path of the fossil sheet from a piece of stone to a microscope table.

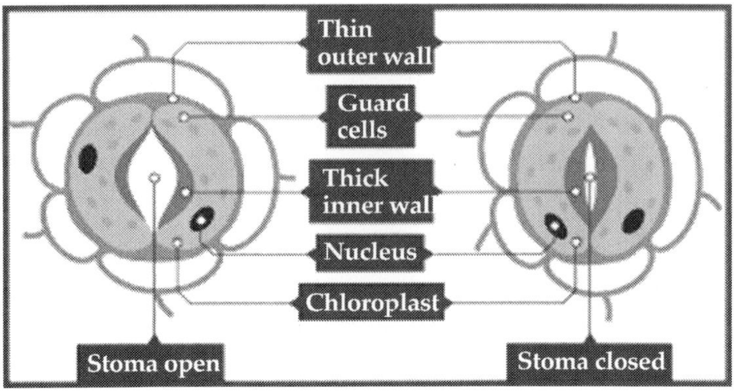

Figure 11: Structure of stomata in plan and cross section Phytolemma in the crucible.

Therefore, on the table is a sample, and on it is a blackened, charred, pressed rest of the sheet. All that remains of the leaf, our known paleobotanist A. N. Krishtofovich called "phytolemic." In Greek, "phyton" is a plant, and "leima" is the remainder. It is not necessary to treat all phyto-leum with acids and alkalis. It is enough to take pieces of it from different places. Sometimes needles and tweezers are enough for this, sometimes it is necessary to treat phytolyme with acid. A black or dark brown slice falls on the bottom of the crucible together with the crystals of Berthelot salt and several drops of nitric acid. Such an oxidizing medium is called the Schulze mixture, paleobotanists have been using it for almost a hundred years. In the Schulze mixture, the phytolemma is from half an hour to several days. It all depends on how much it has changed since the time it was buried in the sediment. Finally, after the bath in distilled water, the phytolemma is poured with ammonia (or another alkali) and is stirred. It dissolves the coal, sandwiched between the cuticles of both sides of the sheet. Under a microscope, you can see how brown clubs creep out of the slit and move the phytolemma.

After a little time passes, in the darkened solution transparent films floats. This is the patchwork of the long-awaited cuticle. With a pipette or a needle, they must be transferred to a slide, spread out, topped with a hot gelatin solution in glycerin and cover with a cover slip. You can look through the microscope.

Everything goes so smoothly, of course, only on paper. Usually you have to show more or less patience. Then the phytolyme does not want to leave the stone, it turns out to be dirty, and it must be etched with hydrofluoric acid, then the pieces of the cuticle are so tiny that they can hardly be moved onto a slide. However, if everything goes perfectly, you can make two or three dozens of drugs a day (Gayangos & Brett, 2002).

NOW FOR THE MICROSCOPE

It is difficult to say with which the paleobotanist meets more often when he finally sits down behind a microscope - with new questions or with long-awaited answers. Before giving answers, the epidermal method requires a lot. It is necessary to understand the complex ornament of cells, stomata and tubercles on the cuticle, which is not always easy. It is necessary to trace how the cellular structure varies from the middle of the leaf to the edge and from the base to the top. It is especially difficult to understand the structure of stomata (there have been cases when researchers took the cells that were behind the stomata for the fallen hairs). Then it is necessary to sort out the literature on the group of plants under study, to see what preparations were obtained from the predecessors. Literature on the structure of the epidermis of fossil plants is already quite large (hundreds of titles), but illustrations are not always good and descriptions are full. Sits paleobotanik, looks at the drug, the picture in the book, reads the description and suffers: did he understand the characteristic features of the species or genus correctly? In addition, if something does not agree, then what is the matter? Maybe a mistake? Then who? Maybe he made a mistake, maybe his predecessor, or maybe it is just that different plants were caught. However, these are already difficulties inherent in any work on the systematics of living beings, and the epidermal method here does not differ from any other.

However, we digress from our simulator plants, from which the conversation about the cuticle began. Let us go back and become acquainted with them. In the Mesozoic deposits, there are often leaves that are a bit like ficus. A thick vein runs from the base to the apex, and thin, lateral veins run from it to the edges almost at right angles, which occasionally branch. The structure is uncomplicated, there are many such leaves, referred to as the genus shadowhopteris (literal translation "ribbon-like fern"), and species were distinguished according to their external appearance. At first, such leaves can be easily systematized: long leaves in one direction, short ones in the other, thick veins - one species, less veins - the other, etc. However, the time passed and the lines between the species became blurred.

At the beginning of this century, the Swedish paleobotanist AG Naggers, one of the patriarchs of the epidermal method, received cuticle preparations from the leaves of the shadowhopteris. Then the work continued Englishmen G. Thomas, N. Bancroft and T. Harris. The results were at least unexpected. Under the monotonous, almost stereotyped leaves, two completely different groups of plants were concealed. One group - the cicadasses - still lives in the tropics, sometimes they are called sago palms. Another group died out with dinosaurs. In honor of the American publicist, organizer of scientific expeditions D. G. Bennett, she was called "Bennettity." This is a complex and largely mysterious group of plants in which the reproductive organs remotely resembled magnolia flowers. At one time, it was assumed that the Bennettites are the direct ancestors of flowering plants, but now they refused this opinion. When they compared the findings of leaves and organs of reproduction in different cicadasses and bennettites, it was found that the second stomata are much more complicated than the first ones, the walls between adjacent cells are sinuous in the Bennettrites, while the cicadas have straight lines. To recognize two such large groups of plants on the leaves, now it is not necessary to wait until the leaves and flowers come across on one branch. It is enough to make a cuticle preparation (Dilcher, 2001).

CONIFEROUS IS ALSO SIMULATORS

Perhaps the most interesting and important discoveries with an epidermal method in hand were made when paleobotanists undertook to study conifers. Most of all worked on them Swedish paleobotanist Rudolf Florin. At the beginning of his work, Florin was engaged in mosses, fossil flowers, and then, on the advice of his mentor, Nathgorst, engaged in the study of coniferous plants. The choice was not accidental.

The remnants of conifers first appear in the Carboniferous sediments and further up the section are found in large quantities in the sediments of all geological periods. The leaves of the conifers are small (in their everyday life they are called the needles) and stick tightly to the branches.

In dealing with other groups of plants, the paleobotanist can rarely enjoy such a coincidence that the leaves are whole, and they sit in large numbers on branches, and are not scattered in the breed singly. This, however, is the only advantage of the remains of conifers. Otherwise, to systematize them in appearance is a real punishment. This is not only difficult, but also unreliable. How deceptive the appearance of conifers can be is seen in Figure 12, which shows similar leaves of different genera (a good example of simulation) and very different leaves of the same genus (Dilcher, 1991).

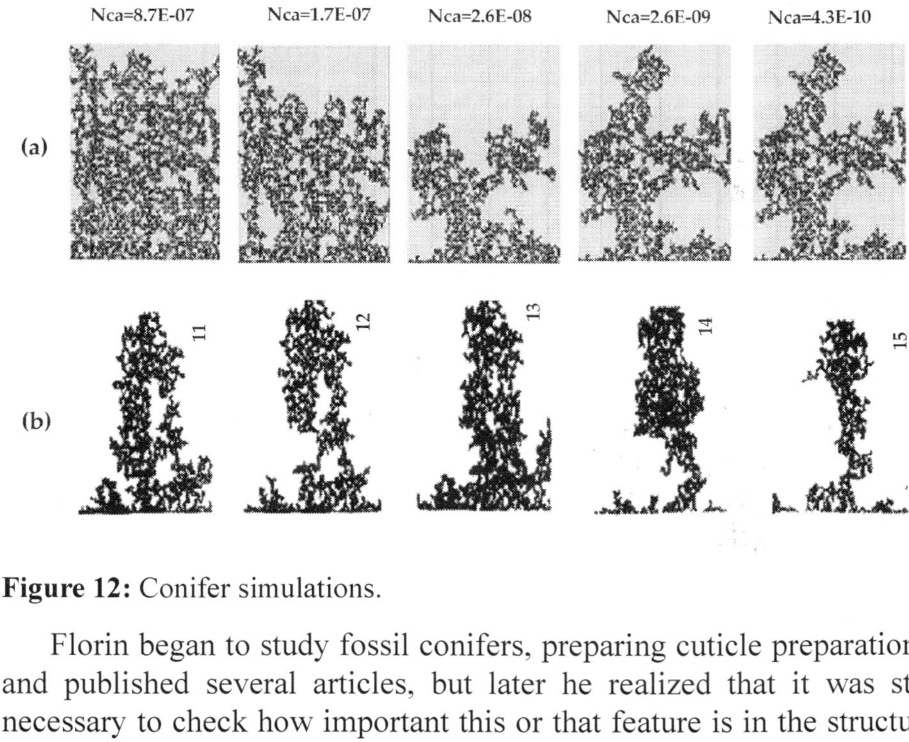

Figure 12: Conifer simulations.

Florin began to study fossil conifers, preparing cuticle preparations, and published several articles, but later he realized that it was still necessary to check how important this or that feature is in the structure of their epidermis, that is, what can be relied upon and what can be neglected. This could only be done with modern conifers, and Florin undertook a gigantic job. He traveled around many botanical gardens, traveled to different countries and collected herbariums everywhere. In each species, he tried to take also the first leaves appearing soon after the germination of the seed, and the old leaves. He cut off needles from trees growing on a swamp and on a dry place. Florin decided to take into account all the possible reasons that could change the microstructure of the leaf. This work took ten years, and in 1931, a thick monograph, a real encyclopedia, was published, in which the results of a huge, unique work were summarized.

By the 1930s, the structure of the cuticle of fossil and modern plants was studied by many specialists who were well aware of the reliability of the epidermal method. However, before the work: Florina no one thought that the microstructure of the sheet can sometimes argue in its importance for taxonomy with such proven, reliable signs as the structure of a flower, cones, etc. From the point of view of general biological considerations, this fact remains unexplained until now. In every plant and animal, there are organs that are carefully guarded by natural selection, even from minor changes. Such, for example, venation of the wing is a lot and fast flying insects. Here, each vein occupies a strictly defined place. The slightest bias will disturb the aerodynamics of the wing, giving advantages to the competitor. In plants, too, it is not accidental that reproductive organs have the greatest constancy. Here, the rate of development is important (otherwise the seeds can mature too late or early), and the constancy of the structure that provides the maximum effect when reproducing offspring. Another thing leaves. Their general outline and venation also have their significance (usually unclear for us), but still small arbitrary deviations are not so terrible, and we observe them quite often. The same deviations were observed in the microstructure. In more arid conditions, the number of stomata grows, and the epidermal cell size decreases.

The reasoning on this score seemed quite logical. Of all organs, the leaf is most in contact with the external environment. Its constant changes are obvious ("then rain, then snow"), so in the sheet you cannot expect much consistency. Nevertheless, Florine's observations showed: yes, indeed, there are some fluctuations in the structure of the epidermis; it is true that both the cell size and the number of stomata can vary even in one species. However, this does not apply to the main features of the epidermis. The structure of the stomata, their location, the ornament of the walls, - the signs are stable and do not depend on random causes (Dilcher, 1991).

Having finished this work, Florin took up fossilized conifers, revised old definitions, described many new species and genera, and showed the patterns of their distribution in space. Florin summed up the theoretical basis for the most accurate and reliable method of studying fossil leaves. This research of the Swedish paleobotanical is not worth the price, and we will return to them when we solve the riddle of pinecone and discuss the drifting continents.

REFERENCES

1. Dilcher, D. L. (1991). The Importance of Anatomy and Whole Plant Reconstructions in Paleobotany. *Current Science*, *61*(9–10), 627–629.
2. Dilcher, D. L. (2001). Paleobotany: Some aspects of non-flowering and flowering plant evolution. *Taxon*. https://doi.org/10.2307/1223702
3. Gayangos, P. de, & Brett, M. (2002). The History of the Mohammedan Dynasties in Spain. *Library*, *2*. Retrieved from http://scholar.google.com/scholar?hl=en&btnG=Search&q=intitle:The+History+of+the+Mohammedan+Dynasties+in+Spain#0%5Cn %5Cthttp://www.archive.org/details/historyofthemoha032395mbp
4. Crane, P. R., Herendeen, P., & Friis, E. M. (2004). Fossils and plant phylogeny. *American Journal of Botany*. https://doi.org/10.3732/ajb.91.10.1683
5. Bradshaw, R. (2007). PALEOBOTANY | Overview. *Encyclopedia of Quaternary Science*, 1567–1574. https://doi.org/10.1016/B0-44-452747-8/00174-5
6. Dal B??, E., Dal B??, P., & Snyder, J. (2009). Political dynasties. *Review of Economic Studies*, *76*(1), 115–142. https://doi.org/10.1111/j.1467-937X.2008.00519.x
7. Manchester, S. R., Chen, Z. D., Lu, A. M., & Uemura, K. (2009). Eastern Asian endemic seed plant genera and their paleogeographic history throughout the Northern Hemisphere. *Journal of Systematics and Evolution*, *47*(1), 1–42. https://doi.org/10.1111/j.1759-6831.2009.00001.x
8. Carcaillet, C. (2007). PALEOBOTANY/Charred Particle Analysis Charred. *Encyclopedia of Quaternary Science*, 1582–1593. https://doi.org/10.1016/B0-44-452747-8/00187-3

3 ON VACATION - IN THE CARBONIFEROUS FOREST

To spend a vacation in the Carboniferous forest, you do not need to go particularly far and do not need a time machine. In the Carboniferous period, forests in our country occupied large areas in the European part, Siberia, Kazakhstan, Central Asia, and the Northeast. Everywhere there were either layers of coals, or simply prints of plants in the rocks. Although in search of such a forest, you will sometimes wade through modern thickets, you can find many interesting things. There were few fans to go to spend free time looking for fossil plants, but some left a noticeable mark in paleobotanical research. Such was the German doctor Hugo Lilienstern, who collected remarkable collections of plants of the Triassic period and he described some of them. In addition, the Englishman, DG Scott, having begun his activity as a railway engineer and having seen plants in coal kidneys later, became one of the best specialists in the anatomy of fossil plants. Together with paleobotanist Oliver, he became famous as a pioneer of "seed" ferns. In life, everything happens, and maybe this book will be the first guide of the enthusiast who went on vacation for fossil plants (Wales, et al., 2013).

Therefore, in the Carboniferous period, i.e. 330-275 million years ago, the forests were in many places on the planet. In the following chapters we will visit Siberia, the Arctic, the Antarctic and other regions, and now we will go to the tropical jungle, without which we would not have the Donbass now, as the Germans had Ruhr, and the Czechs and Poles had Silesia.

The Carboniferous period has not received this name in vain. Although in some subsequent periods of coal accumulated not less, and even much more, it is in this! Coal deposits were formed, which are most intensively developed by humankind. At that time, for the first time in the history of the Earth (the coals of the previous Devonian, the period is rare), the situation was exceptionally favorable for the accumulation of plant mass: there was a warm and humid climate, lush vegetation, a suitable relief. One of the largest coal accumulation belts passed through Europe from Ireland to Donbass, Turkey and the North Caucasus at that time (Fig. 13).

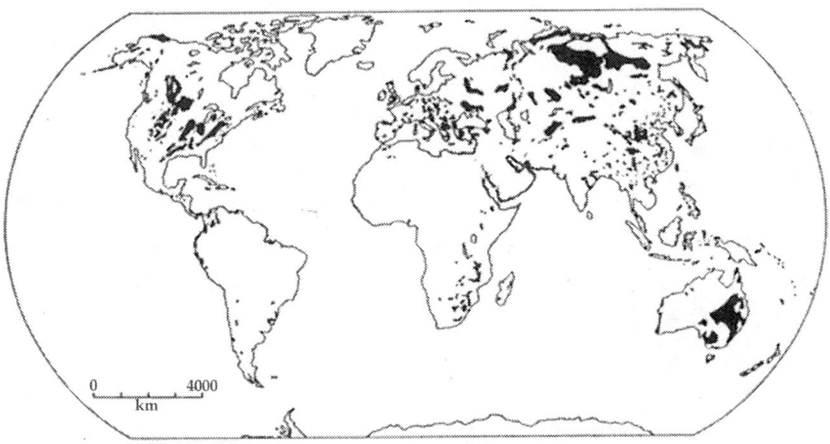

Figure 13: Distribution in Europe of Coal Basins.

The history of Donbass in the Carboniferous period has been studied in great detail, but we will not go into it now. We will only say that during the whole period the sea came here and a few dozen times. Therefore, the geological section of the Donbas is a monotonous alternation of marine and continental rocks: limestones, sandstones, shales, and coals. The latter were formed on extensive swampy coastal lowlands. Plant prints are very common, and in their study our palaeobotanists MD Zalessky, EO Novik and others have invested a lot of work. Donetsk flora is the richest of its peers in our country. Let us see its main attractions (Wales, et al., 2013).

DINOSAUR PLANTS

Fossil plants do not carry on Russian names. This is understandable. In the people they are given names very rarely, and experts prefer Latin. One of the first Russian paleobotanists, Yakim Zembnitsky tried to accustom his colleagues to Russian names, but his attempt was unsuccessful. His "veins" (neuroptis), "ruptishshcha" (schizaea) and "sclerophyllum" (lepidodendron) did not take root. The last word in a slightly modified form of "scheuhedreev" is no-no, and will flash in popular literature when it comes to lepidodendrons. In Greek, "lepis" is scales, and "dendron" is a tree.

Lepidodendron (Figure 14) belongs to an extinct group of plants, from which only the herbaceous plains survived (described in chapter XI) and some exotic genera. Among the plants of coaliferous forests, lepidodendrons were perhaps the largest: up to 30 m in height only up to the first branching of the trunk (a ten-story house) and up to 2 m in diameter near the lump. If you have read about giant mammoth trees, you will not be surprised by such dimensions. However, in all modern forest giants, the trunk consists mainly of strong wood, and the core and bark take up very little space. In lepidodendrons, everything was the opposite: a narrow ring of wood was lost in a continuous mass of a complexly arranged crust. The appearance of the tree as a whole was also unusual. Column-like trunk high up at the top divided into two. Each of the branches was again divided in half, and so it went to the smallest branches. We are accustomed to seeing on our trees that from the thick bough in the direction of more delicate branches. Lepidodendron did so only in rare cases (Tiffney, et al., 2001).

Figure 14: Lepidodendron (reconstruction of MD Zalessky.

On the upper branches of the giant were long and narrow leaves. Leaves almost a meter long with a width less than a centimeter were found. To match the trunk were also impressively sized bumps. Some of them were up to half a meter. Like all plumaceous, lepidodendrons multiplied by spores, which ripened in each cone several tens of thousands. It is believed that the trunks of these trees and their nearest relatives, accumulating in swamps, gave rise to powerful coal seams, although, of course, there were other coal producers (Tiffney, et al., 2001).

Lepidodendron is one of the best-studied fossil plants. We know all its parts, although we find them separately. However, they were not only found separately, but also called differently. It is difficult to understand such manipulation with names at once to paleobotany. Why should the tail of a dog be called a Beetle, paws - Trezor, and ears - Ball? The explanation for this is quite simple. We see the dog in the street entirely, and the paws do not run separately from the tail, there is no sense in giving them independent names. With fossil plants (lepidodendron is no exception), the situation is different.

There was a tree. On it grew leaves, which then fell off, the cone ripened and also fell off, and the wind broke off a branch and carried it to the side. Then the barrel itself collapsed, growing old. There remained only a stump with roots. In such a "disassembled form" interspersed with parts of other trees and lay down lepidodendron in the rock to our days. Before all the parts can be assembled (such luck rarely happens in paleobotany), we must somehow call them. That is called: the trunk - lepidodendron, cone - lepidostrobus, the root part - stigmaria, etc (Besley & Reynal-Querol, 2013).

However, even in those cases when the reconstruction of the general appearance of the plant was successful, all these names cannot be canceled. Here is an example: both the lepidodendron and his close relative of the sigillaria (Zembnitsky called her "typist") had the same root system (stigmaria). However, it is unreasonable to call the remainder of the plant the same in appearance to call all these names at once. It is better to choose one, independent. In addition, they do.

TREE-LIKE HORSETAILS

If you are familiar with marsh horsetails, mentally increase it once in 20-30, and it will look like a calamite (Figure 15), a characteristic plant of coal. To supplement the similarity, the horsetail will have to change something. The trunks of the calamites were thicker, and much stronger. These were real trees with well-developed strong wood, the layer of which in the trunk reached 12 cm on each side. The wood of the calamites was in many respects similar to the wood of the Paleozoic conifers. In horsetails there is no such wood. Their leaves were also different. In horsetails there are small scales that sit round in knots, in calamites the leaves are better developed, although they were never very large. The prints of the leafy branches of the calamites are very decorative. However, the most important difference between horsetails and calamites is in the reproductive organs. The first sporangia (i.e., pouches with spores) are collected in neat little cones, with which the stalk ends, the spores themselves are the same and have long strands wound into the springs. When the spores mature, the springs unwind; under such sails, spores fly in the wind far from the parent plant. In most calamites, the spores apparently were without springs, and cones with sporangia in large numbers sat among the branches (Besley & Reynal-Querol, 2013).

Figure 15: Calamint.

Like lepidodendrons, calamites never fall into the hands of paleobotanics entirely. When the plant died, its trunk was easily broken in knots. The wide cavity inside the trunk quickly filled with silt or sand, which then hardened. Impression of the inner part of the trunk was obtained. They are often found in Paleozoic deposits, but are usually useless for both the geologist and the paleobotanical (McCourt, et al., 2004).

The distant relatives of the calamites were small-sized herbaceous spheno-phylla, which Zempnitsky successfully called "clinolists" (in Greek "sphene" - wedge, "phyllon" - leaf), and this word is found in our literature. Here the leaves were short, greatly expanding to the apex. Sphenophiles in the Paleozoic lived throughout the world and died out at the beginning of the Mesozoic era. Among them, forms with hooks on stems were found. It is believed that some of the sphenophiles were climbing plants.

ABOUT THE FERNS OF GENUINE AND IMAGINARY

In the Carboniferous forest, there were many ferns, some of which belonged to families that survived to this day. Many had large, columnar trunks. Such tree-like ferns can still be found in the tropics (McCourt, et al., 2004). The fossilized trunks of Paleozoic tree ferns were first described by a German scientist called Saarsteine, which means "starling stone." Indeed, the transverse section of these ferns resembles the pattern on the chest of the starling. Polished pieces of "starling stone" are used for small crafts.

Studying fossil ferns is a difficult task. Outwardly, their impressions are very effective, but paleobotany admires this beauty only until they begin to establish genera and species. It is very difficult to rely on the prints of leaves deprived of sporangia. Of course, and for such specimens, you can choose generic and species names, but this paleobotany is not enough. It is important to determine to which family the fern belongs. For this, it is necessary to know the structure and location of sporangia (Figure 16). To botany to consider everything in modern ferns, it is enough for him to take a microscope or even a strong magnifying glass. Paleobotany has to use complex methods. Sporangia on fossil leaves are usually squeezed, pressed against a sheet, the intervals between them are clogged with sandy or clay particles. Sometimes the sporangia are completely covered with rocks and only slightly visible through the print. Even in a microscope, it is not always possible to discern the number of sporangia, their location, shape, cellular structure. Without all this, it is impossible to determine whether a fern belongs to a particular family (Wilf, et al., 2005).

Figure 16: Fern leaf with sporangia from the Carboniferous deposits.

A simple way out of the situation was suggested again by John Walton - the one who came up with a way to get rid of thin sections when studying coal kidneys. His method is as follows. A specimen with a spore-bearing fern is glued to the glass with a Canadian balsam and covered with paraffin. Then the paraffin is scraped off the back of the sample (Wilf, et al., 2005). All “construction” is immersed in hydrofluoric acid, which begins to corrode the stone and, finally, dissolves it. The paraffin-treated glass does not affect the acid. Now on the glass there are only the clean, charred feathers of the fern with the sporangia attached to it, which can be viewed from all sides. Particularly clearly visible are all the details in photographs taken in infrared light.

From the ferns of the present we pass to the ferns imaginary. It is not hard to imagine what a sensation would have caused a find of cucumbers grown on pine. Perhaps, not such, but close feelings were experienced by paleobotanists at the beginning of this century. The history of the discovery, which will now be discussed, goes back to the beginning of the last century. Paleobotany took the first steps, but in the Carboniferous sediments of Europe, rich collections of fossil plants were already collected. In France, they were then engaged in the father of modern paleobotany Adolf Bronyar. Systematizing fossil ferns, he drew attention to the fact that many of them never encounter sporangia. To attribute such remnants to natural groups, for example families, was impossible, and Bronyar was the first in the history of paleobotany to introduce formal systematics, in this case for ferns. The word “formal” here means that the systematics is not based on related plant connections (which are simply unknown), but on formal characteristics chosen from the point of view of convenience (Wales, et al., 2013).

Paleobotany widely used (and now apply) the formal classification of Bronyar, built very successfully. However, at heart they were sure that they were dealing with real ferns, only an unknown kinship. Only at the end of the last century the systematic absence of sporangia on leaves of a certain type-aroused suspicion that some of the coal ferns belong to some other group of plants. This was also evidenced by the results of anatomical studies of some species. In 1898, the French paleobotanist René Zeiller said with great hesitation about such plants that these are “cicadas with fern foliage.” It seemed that paleobotanical thought was ready to part with old prejudices.

Yet published in 1903 and 1904, the articles of F. Oliver and D. G. Scott produced a sensation. These English paleobotanists described the typical ferns on which the real seeds grew (Figure 17). Even many years later one of the authors of the discovery, Scott, wrote about these plants: "We still cannot get rid of the fact that they are ferns - so they look like them." A. Seward wrote about them: "This is a product of evolution, as if purposely created to serve as a warning to botanists, who often assume the outer similarity for an organic relationship."

Figure 17: Part of the leaf 'seed fern' (Neuropteris); A large seed is on the top.

In the early years of this century, popular publications have written a lot about the discovery of Oliver and Scott. It turned out, as always after some new biological observation. Soon, many similar cases were discovered, by which they simply passed. Paleobotany Grand-Erie, Arber, White, Kidston, a little later Halle and others found the seeds on other "ferns." Since then, the number of such finds has increased even more. Now in each course paleobotanists and botanists mention seed ferns (Wales, et al., 2013).

Strictly speaking, the name "seed ferns" is unsuccessful, since it suggests some ferns, which in the course of evolution gained the ability to multiply by seeds. In fact, both real and seed ferns are two independent lines that separated from the as yet unknown common ancestor in the Devonian. Each group independently produced similar foliage. This fact in itself is no less amazing than if we had met seeds on the leaves of real ferns.

OTHER PLACES OF INTEREST

Acquaintance with lepidodendrons, calamites, sfeno-phyllis, ferns and pteridosperms, apparently, was rather tiring, and the rest of the sights we will consider at a faster pace. It is not that other Carboniferous plants are less interesting. In each of them there are peculiar features that have brought many mysteries to researchers. However, it is impossible to tell about everything in detail.

In the Carboniferous forest, in addition to the real and seminal ferns, there were also "primal ferns." What place they occupy in the phylogenetic tree is still unclear. Maybe they are the ancestors of all the ferns. These include biskaliteka, which was discussed in the section on coal kidneys. A characteristic feature of this group of plants was a peculiar branching of the leaves. Strictly speaking, many of them had no leaves, and rounded or whimsically branching flattened bare axes protruded in all directions.

The last group of plants, which is worth mentioning, is the Cordaites, the possible ancestors of our conifers (Tiffney, et al., 2001). They were tall, slender trees with beautiful leaves, between which were earrings bearing either seeds or sporangia with pollen (Figure 18). It was long believed that the Cordayites occupied drier places in the Carboniferous forest. This opinion about the lifestyle of the Cordayites turned out to be only partly true. Recent anatomical studies of the roots of the Cordayites found in them air-bearing cavities peculiar to plants of mangrove thickets, i.e. thickets on lowlands, periodically flooded by the sea. In them cordaites grew with lepidodendrons.

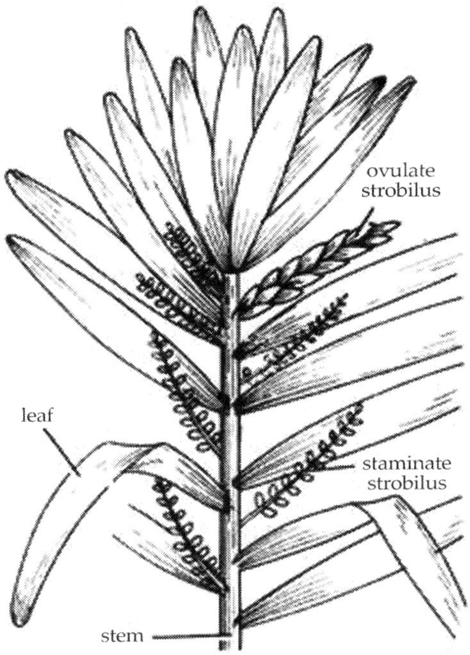

Figure 18: The branch of the European cordaite.

The leaves of the Cordayites are interesting from a mechanical point of view (Tiffney, et al., 2001). They were quite large (up to a meter in length), ribbon-shaped, slightly reminiscent of leaves of iris or corn. Numerous almost parallel veins ran along the sheet. Between them was a system of longitudinal beams formed by strong fibers of mechanical tissue. It was a very sturdy construction. With the Cordayites we will meet again when we are acquainted with the fossil rifts of Siberia ("cordayite taiga"), discuss the drifting continents and the origin of pinecones.

WHY DO YOU NEED 400,000 PRINTS?

In the 1920s, articles by D. Davis, who studied fossil Carboniferous plants in South Wales, were published. On a site of several tens of square kilometers and 700 meters of a geological section, he collected nearly 400,000 prints, tying all the finds to 29 coal seams. After determining the prints, scrupulous counting began, which plants are encountered and which ones are avoided (Taylor, et al., 2009). The results were shown in graphs. Davis came to the conclusion that the lepidodendrons and their closest relatives lived on the damp ones, and the real and seminal ferns lived in drier places. Data were obtained on the way of life and other plants. Was it worth it to start a truly hellish labor?

Since the end of the last century, in popular publications began to appear reconstruction of landscapes of past geological epochs. Drawing the attention of the reader, these pictures at the same time forced the paleobotanist composing them to ponder over the general appearance of plants with the remains of which he worked, to present them in a natural setting. I must admit that this very naturalness of the situation on the majority of reconstructions left much to be desired. Rather, you might think that the picture depicts a fantastic botanical garden with plants neatly arranged for viewing.

It was not the fault, but the trouble of the paleobotanists, at whose disposal there were almost no corresponding observations. All the time, the study of the fossil plants itself was taken away. It was necessary to develop their taxonomy, to determine the numerous collections, to solve the most pressing issues of the geology of the coal basins. Only a few paleobotanists paid attention to the way of life of extinct plants. Such observations can be found already in fairly early works, but the researchers were mainly based on the structure of plants, and not on the analysis of their "cemeteries." (Taylor, et al., 2009)

When it comes to the rough dating of rocks, it is usually sufficient to establish a common set of genera and species, not paying special attention to what companies and what kind of rocks they meet. However, increasing the detail of geological studies, the need to restore the conditions in which rocks were deposited, attracted attention to the burial of fossil plants. The first systematic observations began in France and Germany at the end of the last century. They largely helped the outstanding German paleobotanist G. Potonier to create the foundations of the doctrine of accumulation of combustible minerals.

It cannot be said that since then a great deal has been done in restoring the way of life of coal-bearing plants. The reasons for this will be discussed in the final chapter. Still important and interesting information was received. Ukrainian paleobotanists AK Shchegolev and OP Fisunenko, for example, recently showed that in the Carboniferous forests of Donbass there were three plant groups. One - with lepidodendrons, calamites, to a lesser extent ferns and cordaites - inhabited the wettest places, coastal marshy lowlands. Another grouping included seed and real ferns and occupied higher places. The most dry habitats were occupied by coniferous and some seed ferns. Thus, the observations of D. Davis were confirmed and detailed. Such detailed studies have made it possible to significantly refine the definition of the age of the rocks, helped to understand the geological structure of the Donets Basin (Taylor, et al., 2009). Systematically paleobotanists are advancing in the study of coal forests. More and more the landscape, separated from us almost 300 million years, comes to life.

REFERENCES

1. Taylor, E. L., Taylor, T. N., & Krings, M. (2009). *Paleobotany. Paleobotany.* Retrieved from http://www.scopus.com/inward/record.url?eid=2-s2.0-77951257305&partnerID=tZOtx3y1
2. Besley, T., & Reynal-Querol, M. (2013). Selection via Dynasties: Theory and Evidence. *Working Paper*.
3. McCourt, R. M., Delwiche, C. F., & Karol, K. G. (2004). Charophyte algae and land plant origins. *Trends in Ecology and Evolution.* https://doi.org/10.1016/j.tree.2004.09.013
4. Wilf, P., Labandeira, C. C., Johnson, K. R., & Cúneo, N. R. (2005). Richness of plant-insect associations in Eocene Patagonia: a legacy for South American biodiversity. *Proceedings of the National Academy of Sciences of the United States of America, 102*(25), 8944–8948. https://doi.org/10.1073/pnas.0500516102
5. Wales, N., Allaby, R., Willerslev, E., & Gilbert, M. T. P. (2013). PALEOBOTANY | Ancient Plant DNA. In *Encyclopedia of Quaternary Science* (pp. 705–715). https://doi.org/10.1016/B978-0-444-53643-3.00202-8
6. Tiffney, B. H., & Manchester, S. R. (2001). The Use of Geological and Paleontological Evidence in Evaluating Plant Phylogeographic Hypotheses in the Northern Hemisphere Tertiary. *International Journal of Plant Sciences, 162*(S6), S3–S17. https://doi.org/10.1086/323880

4 DIFFERENT PLANT WORLD IN SIBERIAN BASIN

Now we will go to Siberia, to Kuzbass. For a person far from geology, and even more so from paleobotany, the difference between the Donbas and Kuzbass lies mainly in the geography and the fact that the Donbas is a veteran among our coal basins, and the years of the prewar five-year plans are associated with the Kuzbass. For paleobotany, both basins are so different that when asking the question about their differences, it is quite possible to count on the counter question: "What do they have in common?" Indeed, the plant world in them was as different as in the modern suburbs and in the tropical jungle (Andornino, 2006).

A HALF-CENTURY DISPUTE OVER AN ERROR

This grandiose dispute ended in the late 1920s. Almost all geologists and paleobotanists who were engaged in or were interested in coal-bearing Paleozoic deposits throughout Siberia took part in it to some extent.

It began with a less important mistake. In the 70s of the last century, the Kiev paleobotanist IF Shmalhausen (the father of the famous evolutionary biologist II Shmalhausen) worked on collections of fossil plants collected in the Kuznetsk, Tunguska and Pechora basins. About the Paleozoic flora of the Kuznetsk basin, almost nothing was known at that time (only two small articles of German paleobotanists, Gainits and Geppert, were devoted to it). The Tunguska and Pechora basins were a paleobotanic white spot. It is possible to understand what interest was evoked by the thorough monograph carried out by I. Shmalhausen, one of the best paleobotanists of the last century. However, it was necessary to happen so that in this perfectly executed for those times monograph, there was an unfortunate mistake. However, Schmalhausen was not to blame for it (Graham, et al., 2001).

Assembling collections of fossil plants (as well as any other geological evidence) must be strictly documented, from which layer samples are taken. To this basic rule the geologist accustoms from the first days of training to field works. In the last century, this was not understood by all researchers, and samples were often laconically labeled. The nearest village, some part of the river, was indicated, and everything was limited. Now such materials a paleobotanik does not take into account at all. A stone without an exact address is suitable only for an amateur collection of rarities. However, in the last century, each specimen brought from Siberia or from Pechora was more interesting for us than Antarctic collections. Therefore, Schmalhausen described all the materials that were at his disposal, including those that were not well documented. He did not know that the samples collected in the Kuzbass were obtained from two essentially different parts of the geological section, namely: from the Permian (Upper Paleozoic) and Jurassic (Mesozoic) deposits. Inattention to the collector and inaccuracy in the label turned into a serious mistake from the geological viewpoint (Graham, et al., 2001).

We already said that the Siberian flora was then almost unknown, therefore Permian plants brought from Kuzbass were new to science, not at all like the plants of the same age in Western Europe. In addition, the Jurassic flora, more monotonous throughout the world, knew quite well. Therefore, the Jurassic prints that appeared in the Kuznetsk collection were familiar to IF Shmalhausen, and this decided the conclusion he had given about his age. His detailed monograph was called “The Jurassic

Flora of Russia." Although the Pechora and Tungus collections typically did not contain Jurassic plants, but the flora of these places was quite of the Kuznetsk type and did not even occur to date it to another Schmalhausen.

Shortly after the publication of the monograph Schmal-gausen, the Russian geologist KL Kosmysky expressed doubts about the correctness of the conclusion about the Jurassic age of the coal-bearing thickness of the Kuzbass. He correctly assumed that there are deposits of several geological systems. Schmalhausen responded to Kosmovsky's speech with a rather harsh article. In the dispute, the prominent French paleobotanist Renée Zeiller joined in. He saw the error of Schmalhausen and based on the collections of the geologist IP Tolmachev, which he had again sent to Russia, came to a conclusion about the Permian age of the Kuznetsk coal-bearing strata. At the beginning of this century, Palaeozoic flora of Siberia began to be occupied by MD Zalessky, later one of the most authoritative palaeobotanists. Zalessky, a faithful disciple of Seyler, unconditionally supported the opinion of his teacher. Until the end of his life, that is, until 1946, MD Zalessky believed that the coal-bearing strata of Siberia were entirely Permian.

It was the other extreme. After all, in the monograph Schmalhausen pointed specifically Mesozoic, namely a Jurassic species (Graham, et al., 2001). How did Zalessky deal with them? It is very simple! In the literature, he found among the known Paleozoic plants at that time about the same appearance in appearance, and Jurassic imprints were attributed to them. Outwardly everything became quite smooth, and on the geological map of Kuzbass, prepared by 1925 by a group of geologists under the leadership of VI Yavorsky and PI Butov, extensive fields of coal-bearing Permian deposits are shown. Jurassic rocks in the Kuzbass did not become. The question was not so much solved, how many are hushed up, although many of the venerable geologists continued to think about it, repeatedly returning to the thoughts expressed by Kosmovsky.

Everything fell into place only in the summer of 1927, when the Tomsk geologist LM Shorokhov, working in the Kuzbass, discovered that the Jurassic and Mesozoic plants lie in different parts of the geological section. A year later, a collaborator of the Geological Museum of the Academy of Sciences of the USSR MF Neiburg, in the future the largest Soviet paleobotanik, came to Kuzbass, not trusting the geologists who collected the fossil plants. On the spot, she began to select the remains of plants, carefully marking the layers in which this or that species was found. She confirmed Shorokhov's conclusions. So the final confusion in the collection, which fell into the hands of Schmalhausen, became finally clear. On the geological map of Kuzbass appeared horbkie deposits. In subsequent years MF Neiburg succeeded in showing that in the Kuzbass there is a complete cut from the Carboniferous to the Jurassic systems inclusive and all this with numerous fossil plants. The vision of Kosmovsky was brilliantly confirmed, at the same time it became a triumph of paleobotany. It is to fossil plants that we are indebted to the fact that at the disposal of geologists who mastered the Kuzbas, there was a reliable geological map (Tomescu, 2016).

Here, however, it is reasonable to ask the question, but was it worthwhile to break the spears for half a century because of the age of coal-bearing deposits with fossil plants? Does it matter how old is coal? He would be good. Unfortunately, such a question is asked by the miners who came to mine coal for the already studied, explored places. When coal combines and jackhammers were used, it is really not necessary to know the age of the layer, but it is impossible to find the deposit, properly to reconnoiter it and prepare it for exploitation, not knowing the age of the rocks. Only having a good geological map with the fields of deposits of different ages shown on it can be traced coal seams hidden from a geologist of a thick stratum of rocks. However, we will return to this issue in the penultimate chapter.

THE KINGDOM OF THE KORDAITES

If a paleobotanical collection of fossil plants from the Permian or Carboniferous (not from the Lower Carboniferous) deposits of the Kuznetsk Basin and in general from Siberia falls into the hands, he cannot be sure, that the specimens will have prints of the Cordait leaves (Figures 19 and 20), though in small numbers, but there. Sometimes their prints so densely fill the surface of the sample, that, when cleaning one sheet for cleaning, it is necessary to spoil a dozen others: a real layered pie from the leaves (Tomescu, 2016). Such accumulations of leaves are sometimes called "forest litter", but this is not true. That soft carpet of leaves, in which mushroom pickers seek the object of their passion, does not go to paleobotany. In the forest, all leaves rot, and from them there is only a trash, as is easily seen by digging in a real forest litter. For a leaf to be preserved for millions of years in the form of a phytolyme or imprint, it must get into the water - a lake, a river, a swamp or the sea. However, the leaf accumulation in the water no one will call a forest litter.

Figure 19: Rosette leaves of the Permian cordierites.

Therefore, the leaves of the Cordayites are as abundant in the coal-bearing Paleozoic deposits of Siberia, as are the needles of conifers in the same places. Therefore, the Cordayites first of all attracted the attention of paleobotanists. Among such frequently occurring fossils, it is most convenient to look for arrows showing geological time (Tomescu, 2016). Unfortunately for the paleobotanists, the leaves of the Siberian cordaid were so much on one face that both Schmalhausen and Zalessky considered it possible to attribute them to the same species. Kordaites became ballast in paleobotanical collections.

Paleobotany, working in Siberia now, it seems paradoxical. In 1964, about a hundred species of Cordaites, naturally distributed along the geological section, already appeared in the literature. However, in the first years of development of paleobotanical deposits of Siberia was not so. There are several reasons for this, but the main thing is that the leaves of the Cordayites were looked in isolation from the geological section. The first questioned was that all Siberian Cordayites were one species, the same MF Neiburg. It was not for nothing that she collected fossil plants on expeditions, and not only sat in the study above the samples. Her casually expressed in a small article of doubt on this account was enough to ensure that the number of species released began to grow rapidly (Tomescu, 2016). By the end of the 30s, the Cordayites not only ceased to be ballast, but also became one of the most important (or, as they often say, "guiding") species. Externally simple and monotonous leaves began to see more and more differences. Paleobotany measured the density of veins, the proportions of the leaf, took into account its dimensions, outlines, distinguished all new and new species and ... overstretched the stick. On the pages of paleobotanical work came the orderly rows of neatly matched by digital indicators of leaves. Intruded into a close mathematical framework, the Cordayites began to take revenge on paleobotanists. It became very difficult to understand the system of leaves that lacked the characteristic of all living variability. Errors in the definitions were becoming more frequent, and incorrect contours appeared on the geological maps.

Figure 20: Leaves of the Angara cordierites.

The history of paleobotany knows many of such strikes of fossil plants (Tomescu, 2016). The way out of difficulties is always one: to penetrate deep into the structure of plants, to revitalize, to make the systematics more natural. When it comes to leaves, we must first try to study the structure of their epidermis. Here we return, thus, to the cells, stomata, cuticles, phytolamps. In a word, again all hope for chemistry, drugs and a microscope, Who has forgotten how paleobotanists disclose the microstructure of fossil leaves, can return to the chapter "Canned food of 300 million years", and we will continue the story of the Siberian cordayites (Tomescu, 2016).

STUBBORN LEAVES

The cellular structure of the Siberian cordierites turned out to be a hard nut to crack. The first attempts to get cuticle preparations and the epidermis by chemical methods failed. I had to adopt another method. It is hard to say who borrowed it from anyone, but only this method is used by paleobotanists, paleozoologists, and ... forensic scientists. When you need to establish which gun the criminal was shooting from, a fatal bullet is subjected to an uncomplicated investigation. It is applied a special liquid (like nail polish), which, after drying, gives a transparent elastic film, repeating all the scratches on the surface of the bullet, left by the irregularities of the trunk. Then from the test pistol they shoot at something soft, with the second bullet they shoot exactly the same transparent copy, and both films are compared under a special microscope. If the prints of scratches are the same, then both bullets are released from one gun. Paleontologists and paleobotanists use a similar method to study the microscopic relief of fossils.

In the early 50s with these films, the Tomsk post-graduate student MD Artamontseva (Parfenova) was approached to the Kuznetsk cordait leaves. She received films (we will call them "replicas" after criminologists) with well-known epidermal cells, but could not reach the diagnostics of species on this basis. The stomata could be clearly seen only on one line. However, an interesting application was made, and M. F. Neiburg became interested in the work of Artamontseva, who was already an influential paleobotanist, a major specialist in geology and fossil plants of the Soviet Union. She instructed me, then a student at the Moscow University, to continue the successful study of Artamontseva (Chuanyou, 1994).

The work was based on collections collected in the Tunguska basin. From the point of view of geological practice, the choice of material is justified. The coal-bearing Paleozoic strata of the Tunguska basin were then intensively studied by numerous expeditions, and without reliable paleobotanical evidence it is impossible to compare small outcroppings scattered in the continuous taiga, to form a clear idea of the geological section as a whole, and to draw a convincing geological map. However, from the point of view of paleobotanists, it was difficult to choose a more unsuccessful district, as neither MF Neiburg, nor her student, suspected.

In the western part of the Tunguska basin (where the bulk of the collections collected by the geologists were from), deposits with thc Cordayites were roasted by intramural volcanic outflows back in the Mesozoic times. All attempts to obtain cuticle preparations from caked coal crusts covering the prints of cordait leaves, turned out to be a failure. Phytolaims lay in the strongest oxidants for weeks, they were cooked for a long time in acids, but they remained the same black. It was necessary to follow the example of Kuzbass to the replicas.

Although the leaf prints had been under pressure for many millions of years, the microscopic (a hundredths of a millimeter) cells were clearly visible on replicas. Of course, the leaves did not surrender immediately, and it took many months for the ordeal to find a mixture for the preparation of replicas, to familiarize themselves with all the vagaries of prints, to collect a fairly representative material.

The work was fine. The replica was enough only after three years, and only then did the whole complexity of the question become clear. Some leaves immediately showed good cells on the surface and stomata scattered here and there, and on others some awkward picture was obtained. Even on the prints of miraculous preservation with perfectly printed cells, the stomata were not visible. However, between the veins, there were impressions of the hillocks located by regular longitudinal rows, on both sides bordering a shallow but very distinct groove (Figure 21). Sometimes on a print there were no traces of veins, and the sheet traced only distinct impressions of these grooves. On specimens of better preservation, it can be seen that such grooves were run one by one between each two veins on one (lower) side of the sheet. Three years have passed, while it was possible to unravel the nature of these grooves. It turned out that these are special receptacles for stomata. Hidden in grooves, locked with tubercles on bordering cells, stomata could not be seen on replicas (Hillman, 1989). All this did not become immediately clear and for a long time remained only an assumption. We needed stronger proofs than replicas, namely: either sections through mineralized leaves with "conserved intestines," or cuticle preparations.

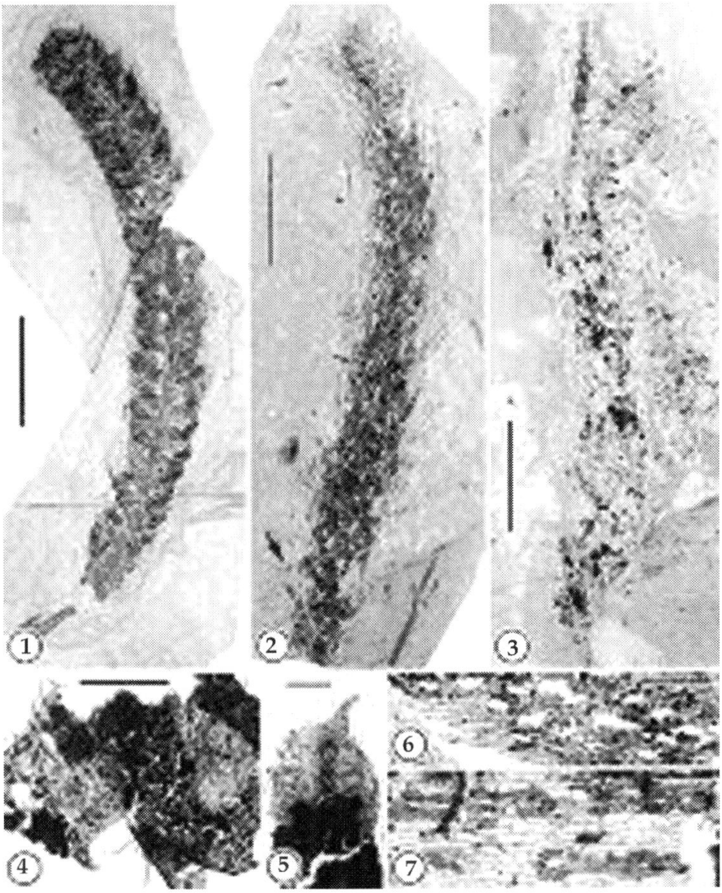

Figure 21: Angara cordait a is impressions of grooves and cells on cellulose.

Finally lucky!

With the discovery of mineralized leaves, everything turned out according to the proverb: "There would be no happiness, but misfortune helped." Ask any of our paleobotanists what he most lacks for happiness, and he will answer you: "Free, convenient cabinets for collections." In the cabinet of M. F. Neiburg, boxes with specimens from all parts of the Soviet Union were constantly arriving. For each new arrival, it was not easy to find a place in the cupboards that had been filled for several decades. Sometimes, it was necessary, reluctantly, to severely clean old collections, throwing out some samples. This is a very painful operation because of the consciousness that it is a shame to throw out a sample that has been waiting for its time for several hundred million years (Andornino, 2006).

Once such an execution was subjected to a collection collected by TA Dobrolyubova on the Pechora River back in the 1920s. MF Neiburg left the most valuable samples, and then, what could be sacrificed, went to the basket. For the first time I attended such an act, every imprint seemed a miracle of nature. My table was not yet cluttered to the limit, and it decided the fate of some samples. They remained secretly, in order to avoid ridicule, lying in the drawer of my desk, although in the descriptions with their numbers the word "thrown out" already stood.

It took several years until these samples surfaced again. In 1962, M. F. Neiburg died tragically. It was necessary to master her richest legacy, to prepare for publication the remaining unpublished monographs, to revise the collections collected over 40 years. They fell on the eyes and "saved" samples. One of them drew attention to the unusual preservation of the Cordaite leaves. Under the binocular loupe it was clearly visible that they were with grooves. However, the main thing is not this, but the fact that the organic crust covering the prints was not thin and black, as usual, but unusually thick and brown. Are they not mineralized leaves? The first section confirmed the conjecture (Graham, et al., 2001).

On the cross-section of the sheet, its entire internal structure was revealed, including the groove sections on the underside. It was possible to prepare and thin sections, passing obliquely along the sheet, almost parallel to its surface (Figure 22). Everything fell into place, but I wanted to look at the stomata of these strange plants. I had to get cuticle preparations.

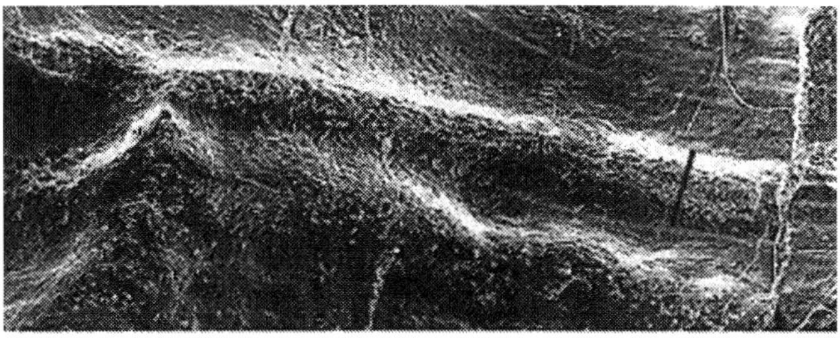

Figure 22: Impressions of the grooves on the print of the sheet of the Angara cordait.

This was done only a year later. In addition, here the role was played not by chance, but by calculation. Geologist and paleobotanist GN Sadovnikov, who studied coaliferous Permian deposits of the Tunguska basin, understood the importance of the works on the systematics of the leaves of the Cordayites, he himself studied them a little and collected their imprints on expeditions on a regular basis (Graham, et al., 2001). He took into account that coal deposits in the eastern parts of the Tunguska basin are much less heated by volcanic outpourings, and decided to look for the material for the preparation of cuticle preparations there. The calculation turned out to be correct, and soon specimens were delivered to Moscow with remarkable safety. The preparation of the preparations did not present any difficulty. Approximately at the same time after viewing several thousand leaves collected in Kuzbass, it was possible to get tolerable preparations from some of them.

Having dealt with the microstructure of the leaves, one could revise their taxonomy. The previous genera and species have been re-examined. Little was left of them after a rigorous audit. It turned out that under one species name there were leaves that belong not only to different species, but even to different genera (Graham, et al., 2001). The old taxonomy, based on bare numbers, has collapsed. The introduction of microscopic methods into the study of the Cordayites soon made it possible to solve a number of controversial questions in the geology of the coal-bearing strata of Siberia.

About other Plants of the Cordayite Taiga

The number of Cordait leaves in the Permian and Carboniferous sediments of Siberia is very high. As paleobotanists say, they make up the main background in plant complexes. Therefore, the opinion of their former dominance in the Siberian landscapes and the existence of the Cordaite taiga in those days are quite understandable. Whether it was true or not, we do not know. Perhaps it was simply the Cordayites who preferred the places around the reservoirs. This could give them significant advantages over their contemporaries in penetrating the pages of the geological record. However, the fact that the Cordayites were notable plants in the Siberian Paleozoic landscapes is undoubtedly.

Otherwise, the population of Siberia was very different from everything we saw in the tropical coal forest of Donbass. In the Kuznetsk and Minusinsk basins, in Tuva, Mongolia, and Eastern Kazakhstan, we find in the Lower Carboniferous deposits a large number of tree-like plaques, but among them there are no lepidodendrons. We see here also plants similar to calamites, but not so large and with different structure of the core. The ridge-carpet landscapes of these places were still devoid of Cordaites (Graham, et al., 2001).

The real domination of the Kordaites began only approximately from the middle of the Carboniferous period. At the same time, arthropod plants appeared (see chapter XIV) with leaves fused from below into a funnel-shaped bowl. Ferns were rare and not at all the same as in the rainforests. Seed ferns also had a very special foliage, which we do not find outside of Siberia. Some of them had round leaves on small petioles and with veins diverging thin veins; in others, leaves were cut into segments, as in a carrot top. About third, long attributed to the genus gondwanidium (Gondwanidium), we will talk when we talk about the drifting continents.

The Permian plant world of Siberia and the Pechora basin seemed already well studied when a new and rich group of plants unexpectedly appeared in it. As early as the 1930s, MD Zalesskii found in the Kuznetsk basin fingerprints similar to conifers, and named them Walchia spinulifolia. Valhia is a kind of European coniferous Permian period. During the Great Patriotic War, the same specimens fell into the hands of MF Neiburg, who worked in Kuzbass on the instructions of the People's Commissar. Even then she suspected that it was not conifers, but mosses, but she did not hurry to resolve the issue until more evidence was obtained.

Purposeful search always brings success, and in the early fifties MF Neiburg picked up a decent collection of supposed mosses from the Permian deposits of the Kuznetsk, Pechora and Tunguska basins. Using the same mixture with which they received replicas (pharmacy collodion) in those years, she took off the phytolum from the prints and studied them under a microscope. The thin ligature of the brown cells in the preparation showed unequivocally: it is leafy stems of mosses (by the way, there are still mosses that are known from the Carboniferous period). This was a complete surprise for the paleobotanists (Smoot & Taylor, 1985). Highly developed leafy-white mosses, some have an almost modern appearance, and suddenly - in Permian sediments. Before that, they were found in Cenozoic and very rarely in older (Jurassic) breeds. In addition, here were hundreds of copies, and even in the Paleozoic. It was necessary to justify everything as it should.

Here very opportunely arrived the article of the American paleobotanist U. Darra with a successful recipe for a mixture that allows you to shoot the phytolyme with the entire print. However, there was a mixture of Darra and one flaw. It includes nitrocellulose, that is, quite ordinary explosives. The path of MF Neiburg to her lay through the permission of the police, disorder with a six-pack package (when several hundred grams were needed) and other usual obstacles in such cases. However, the mixture paid off all the trouble. Thanks to her, MF Neiburg obtained excellent preparations (Fig. 23).

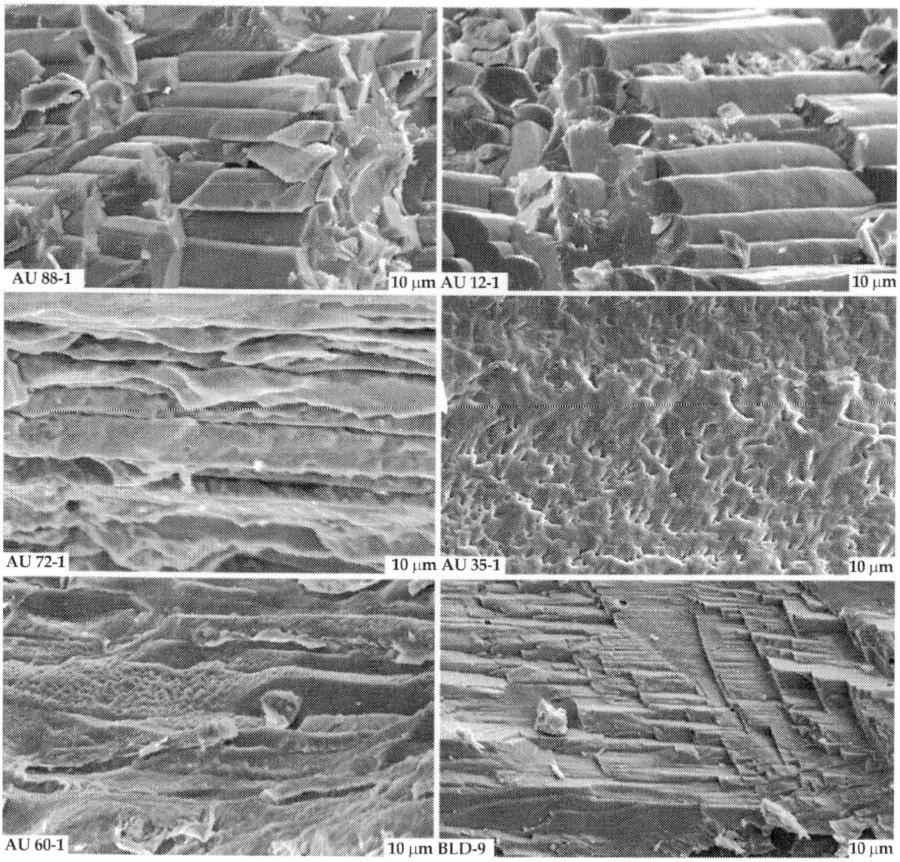

Figure 23: Angara Permian mosses and their microstructure.

In 1956 a small article was published, and in 1960 there was already a large monograph describing 10 genera and 14 species of Permian mosses. Having received from Moscow a parcel with this monograph, a member of the English Royal Society, TM M. Harris (described in chapter X), wrote to M. F. Neiburg in a reply letter: "With one work, you changed the existing ideas in paleobotany, which rarely falls to the lot Scientist ." We did not talk about many other sights of the Siberian corday taiga (Smoot & Taylor, 1985). Indeed, in the Carboniferous and Permian sediments of Siberia, several hundred species have now been established, and in each there is something interesting and unusual. However, with some plants we will meet in the following chapters.

REFERENCES

1. Tomescu, A. M. F. (2016). Development: Paleobotany at the High Table of Evo???Devo. *Current Biology*. https://doi.org/10.1016/j.cub.2016.05.001
2. Chuanyou, G. (1994). Textual research on the history of bamboo garden in China. *Journal of Bamboo Research, 3*. https://doi.org/CNKI:SUN:ZZYJ.0.1994-03-009
3. Hillman, G. C. (1989). Late Palaeolithic Plant Foods from Wadi Kubbaniya in Upper Egypt: dietary diversity, infant weaning, and seasonality in a riverine environment. In *Foraging and Farming: the evolution of plant exploitation* (pp. 205–233).
4. Andornino, G. (2006). The Nature and Linkages of China's Tributary System under the Ming and Qing Dynasties. *Working Papers of the Global Economic History Network*, (21).
5. Graham, A., Gregory-Wodzicki, K. M., & Wright, K. L. (2001). STUDIES IN NEOTROPICAL PALEOBOTANY. XV. AMIO-PLIOCENE PALYNOFLORA FROM THE EASTERN CORDILLERA,BOLIVIA: IMPLICATIONS FOR THE UPLIFT HISTORY OF THE CENTRAL ANDES. *American Journal of Botany*, *88*(9), 1545–1557.
6. Graham, A., Gregory-Wodzicki, K. M., & Wright, K. L. (2001). Studies in neotropical paleobotany. XV. A Mio-Pliocene palynoflora from the Eastern Cordillera, Bolivia: Implications for the uplift history of the Central Andes. *American Journal of Botany*, *88*(9), 1545–1557. https://doi.org/10.2307/3558398
7. Smoot, E. L., & Taylor, T. N. (1985). Paleobotany: Recent developments and future research directions. *Palaeogeography, Palaeoclimatology, Palaeoecology*, *50*(1), 149–162. https://doi.org/10.1016/S0031-0182(85)80009-2

5 WAS THERE A CONTINENT OF GONDWANA?

The appearance of some fundamental geological views was affected by the direct influence of paleobotany. This is how it happened with the concept of "Gondwana." It is a hypothetical continent that once united Africa, South America, Australia, India and Antarctica. Here, in fact, we should say: "apparently, uniting." Although Gondwana has many studies devoted to it, international symposiums are regularly held, and some researchers simply do not think of the former geological epochs without Gondwana, yet the continent remains hypothetical.

On this issue, there are three points of view. Some geologists believe that the above-mentioned Gondwana continents were once connected to Gondwana, and then broke away from each other and moved along the surface of the Earth, leaving a place for oceanic depressions (we will talk about the hypothesis of continental drift in another chapter). Other geologists believe that these continents have never been unified and that the oceans that separate them have always existed, that is, Gondwana as a single continent has never existed. Finally, a third believes that the Gondwana continents, although they remained in their places, were united in a giant continent along continuous massifs of land. The latter then split and its internal parts sank, forming a bed of oceanic depressions (Wilf, & Escapa, 2015).

There are questions in science that are sometimes called "damned." Many theories are put forward on them, each is confirmed by facts, but each has some contradictions. True, sometimes these contradictions are reduced to a simple thesis put forward by one side: "We cannot imagine that ...." The Gondwana issue is one of the veterans, it will soon be a hundred years old. Geologists and geophysicists, zoologists and botanists, geographers and oceanographers continue to argue about it. In addition, the final decision does not come about anyway, partly, perhaps, because the researchers are more likely to relish the contradictions of their opponents than pay attention to their own. At the beginning of the chapter it was said that the birth of the concept of "Gondwana" paleobotany had a direct relationship (Taylor, et al., 2009). However, can't it put forth a weighty word in the dispute and now?

CAPTAIN SCOTT'S DISCOVERY

The beginning of this century was marked by great events in the study of the polar countries. In 1909 Robert Peary successfully reached the North Pole. A year later, two expeditions immediately began to storm the South Pole. Members of the British Antarctic Expedition under the command of Captain Scott came to the pole already when there was a sign put by Amundsen. This was not the only failure of the British. Both Scott himself and his comrades died on the way back. Their last parking was managed to find, and the scientific materials collected by the expedition, fortunately, were not lost. Among the geological samples were plant prints. From the diary it followed that the members of the expedition Wilson and Bowers found them on Mount Buckley, i.e., on the 85th. W., 450 km from the pole. They collected 14 kg of coal and shale samples with plants, despite the fact that they were on the verge of dying from exhaustion. Only a year later the samples were delivered to England. The largest English paleobotanist A.C. Seword described a unique collection in 1914. Although the samples were not very good preserved, there was no doubt that in the Permian time the Antarctic was inhabited by a "glossopteriferous" flora (He, 2006).

Now no one was surprised, because some people had known about this for a long time, and others do not know what the glossopteria flora is. However, when the specimens of Scott's expedition were delivered to Europe, everyone was amazed - both knowledgeable and ignorant. It was astonishing that the prints were found in the very depths of Antarctica, where now there is nothing alive. In addition, not some dystrophic, depressed plants were found, but full-scale prints of normal plants, exactly the same ones that were known in ... India, South Africa, Australia and South America (Taylor, et al., 2009).

After Scott, several expeditions visited the Antarctic that brought Paleozoic fossil plants (Figure 24). Most of the material was collected by the participants of the Transantarctic expedition of 1955-1958, organized by the British. Honor to process this collection fell paleobotanic from the University of Witwatersrand (South Africa) Edna Plumsted. In 1962 she published a detailed monograph in which she summarized everything that we know about the extinct plants of Antarctica. In this capital monograph 140 pages of text and 28 tables with photographs of plant remains. Over 40 plant species have been found in the Paleozoic deposits of Antarctica. Much for the "ice continent." However, it is time to be acquainted with these plants.

Figure 24: Locations of Late Paleozoic Plants in Antarctica.

For such a boring title, one has to apologize to the readers, but it is not chosen by chance. The concept of a "glossopterian flora" is so essential for all our further reasoning that one must get accustomed to it. To better attract the attention of the reader, I put this concept in the headline.

At the beginning of the last century, the French paleobotanist Adolf Bronyar received a small collection of fossil plants from India.

Among them were original and then still unknown leaves. They were linguistically prominent with a thick vein, crossing the leaf from the base to the top. Thin lateral veins joined by bridges formed a complex grid on the sheet (Figure 25). For the tongue-shaped form these leaves were called Glossopteris (in Greek "glossa" is the language, "pteris" is the root word, which is often added to the names of modern and fossil ferns; Bronyar thought that before him there were leaves of some fern) (Escapa, & Cúneo, 2005). In the following decades, glossopterites were found in large numbers in Australia, Africa and South America (Figure 26). It became clear that the paleobotanists encountered a still unknown flora, once occupying large areas. Interestingly, in 1775 Capt. James Cook, returning from a round-the-world trip, finally proved the incorrectness of the assumption of the existence of a certain Southern Earth (Terra Australis). Exactly one hundred years later, in 1875, GF Blanford first drew attention to the spread of glossopteris in India, South Africa and Australia, introduced the concept of "glossopterian flora" and suggested that from the beginning of the Permian period and until the middle of the Tertiary period there was Indo-Ocean continent. Later M. Neymar suggested the existence of the Brazilian-Ethiopian continent. Finally, in 1885, the famous geologist Eduard Suess, the author of the amazingly complete and deep coverage of the material of the "Face of the Earth", proposed the term "continent of Gondwana." (Taylor, et al., 2009)

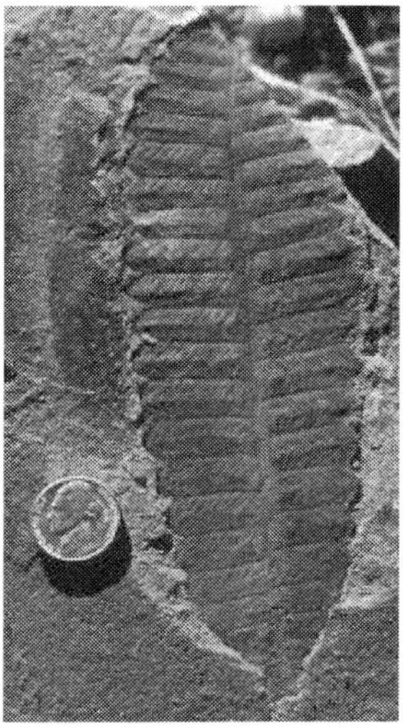

Figure 25: Sheet of glossopteris.

Paleobotany has been around long enough to puzzle over the age of the glossopterina flora (Escapa, & Cúneo, 2005). As in the case of the Kuzbass, at first the Jurassic age was assumed. In the Jurassic deposits of Europe by that time, plants were very similar to glossopteris, but the leaves were not single, but gathered on one stalk for four to five pieces. The relationship of these European leaves and glossopteris seemed beyond doubt (then it turned out to be a mistake). Only when in India and Australia, together with glossopterites, the leaves of the Cordaites were found, it became clear that the glossopterina flora was Paleozoic.

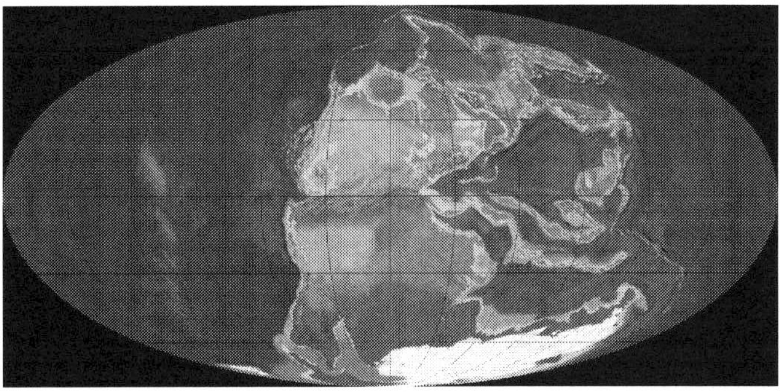

Figure 26: Climatic zones at the boundary of the Carboniferous and Permian periods.

In solving the question of the age of the Kuznetsk coal-bearing layer, the error played a fatal role. With Gondwana coal deposits, something similar happened. In 1883, the French paleobotanist Renée Zeiller described a small collection of fossil plants collected in the Gondwana deposits of the Zambezi River (South Africa) (Lee, 2002). The conclusion of Seyler was completely unexpected: in the collection were only species well known in the Upper Carboniferous sediments of France. Zeiller was already known as a good connoisseur of European Paleozoic plants, and no one doubted the correctness of his definitions. In Africa they also knew the typical Gondwanian glossopteria flora. About the Mesozoic age, almost no one spoke about it, and the question of the age of the glossopterian flora was decided by the method of elimination: if it is not Carboniferous and not Mesozoic, it remains to be attributed to the Permian period. Then with this thought it was necessary only to get used to - and go on and prove that at the very beginning a mistake has been made. In addition, such a mistake was, and very annoying.

Zeiller's article on plants from the Zambezi occupied the minds of paleobotanists. In 1951, the Portuguese paleobotanist K. Teixeira spoke at an international congress with a special explanation of the unusual penetration of Carboniferous European plants far to the south of Africa, with conclusions about the time of the appearance of the glossopteria flora. It would be better for Teixeira not to make his report, because at the same congress the Dutch paleobotanist W. Jongmans told the audience that Zeiller described not African, but typical French flora: someone once confused labels on cabinet drawers.

Although the Gondwana plants have been studied for many years, the dispute about their age is far from over. We know that they are Paleozoic, but in our time this is too little. It is necessary to know at least to what period they belong. Now no one says that they are Mesozoic, they only argue about whether there are coal plants here or only Permian ones. The main reason for the protracted discussion is the peculiar appearance of the Gondwana flora, its striking difference from the flora of the Northern Hemisphere, where geologists established the main geological periods.

Glossopteris and the closely related genus gangamopteris, whose leaves were exactly the same, but without a thick vein in the middle, are as frequent in the Paleozoic coal-bearing deposits of Gondwana as the leaves of the Cordaites in Siberia (Lee, 2002). They also massacred the breed, so it turns out "puff cake from the leaves." They meet, though not too often, other plants. These are horsetails with fused leaves, cordaites, seeds, rare conifers, ferns and various exotic strange plants.

MANY CONTINENTS - ONE FLORA

If the botanist chose for himself the theme "Comparison of the flora of Antarctica and India", he was hardly considered quite normal. Does it make sense, really, to compare the tropical jungle with a miserable set of mosses and lichens of Antarctica? Exactly the same question, when it is put for late Paleozoic time, is very relevant. Three quarters of Paleozoic plants known in Antarctica are also found in India. Of the forty-small Antarctic species, only two or three are unknown on the other Gondwana continents. Approximately the flora and other regions of Gondwana are similar (Taylor, et al., 2009). The further the paleobotanists work, the more fully this extraordinary similarity, the underestimation of which in the past was often associated with incorrect plant definitions or with insufficiently representative collections. Of course, against the background of such similarities, there are differences. However, without this the picture would be just fantastic. Importantly, we do not see any regular changes in the composition of the Paleozoic flora, when we move from one part of Gondwana to another. Such regular changes are characteristic for large areas of the Northern Hemisphere. This can be seen from the example of the Corday taiga. For example, in the Tunguska basin we do not find many Kuznetsk plants, and then again we meet them in Norilsk and in the Western Taimyr. Here such differences are certainly not related to oceanic depressions. Even stranger is the unity of the Gondwana flora, the location of which is now divided by huge oceans.

In the unity of flora of the Gondwana continents there is only one crack. However, this exception, rather, confirms the rule. We are talking about this. In Southern Rhodesia, Argentina and South Brazil, together with the usual glossopteria flora, ferns and clenolists are found in small numbers that are not characteristic of other Gondwana continents, but they are found in Europe, North America, China and Korea. Although these aliens are found only in a small part of the section, their presence indicates the similarity of the continental flora, now separated by the ocean.

The unity of the flora of the Gondwana continents continued in the Mesozoic, but gradually each continent acquired an increasingly distinct identity. Now Africa and India belong to the Palaeotropic floristic region, South America to the Neotropical, and Australia to its own Australian region. Still many plants still guard the ancient traditions that have developed in the Paleozoic and remind us of the former unity of the Gondwana continents. Thus, the genus with the original name "symphony" is found only in the west of Central Africa and in the north of South America. The Proteaceae family accurately depicts Gondwapus (Figure 27), and is found almost throughout South America, in southern Africa, in Australia, India and Madagascar. Only in Antarctica now do not live proteaceous, for a reason they do not have to wait (Lee, 2002).

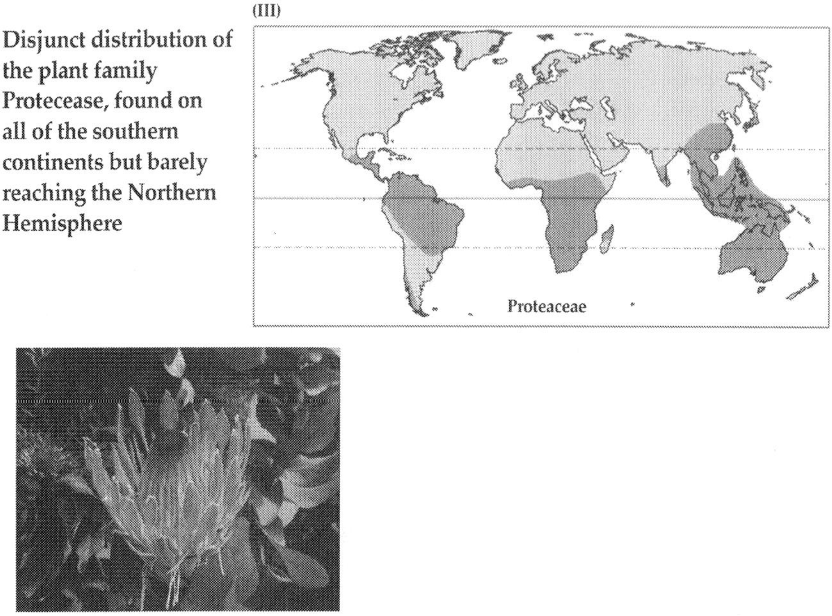

Figure 27: The modern distribution of the family of Proteaceae.

All that has been said is quite enough to understand the feelings of paleobotanists, when they resolutely oppose attempts to lay 1000-kilometer oceans between the Gondwana continents already in the Paleozoic. Gondwana was one for them, but whether the constituent continents were sitting on their places, connected by bridges, or they were floating on the surface of the Earth - which is another question.

REFERENCES

1. Taylor, E. L. ., Taylor, T. N. ., & Krings, M. . (2009). *Paleobotany. Paleobotany*. Retrieved from https://www.scopus.com/inward/record.uri?eid=2-s2.0-77951257305&partnerID=40&md5=2215683d7eba4e99e11abfe624c01477
2. Wilf, P., & Escapa, I. H. (2015). Green Web or megabiased clock? Plant fossils from Gondwanan Patagonia speak on evolutionary radiations. *New Phytologist*, *207*(2), 283–290. https://doi.org/10.1111/nph.13114

3. He, D. (2006). The expansion of water communication and transportation between north and south China during the Six dynasties. *Frontiers of History in China*, *1*(2), 236–253. https://doi.org/10.1007/s11462-006-0004-2
4. Escapa, I., & Cúneo, R. (2005). A new equisetalean plant from the early Permian of Patagonia, Argentina. *Review of Palaeobotany and Palynology*, *137*(1–2), 1–14. https://doi.org/10.1016/j.revpalbo.2005.06.002
5. Lee, Y. K. (2002). Building the Chronology of Early Chinese History. *Asian Perspectives*, *41*(1), 15–42. https://doi.org/10.1353/asi.2002.0006

6 THERMOMETERS OF THE GEOLOGICAL PAST

The weather tomorrow or, for example, in the coming year interests everyone. The weather yesterday or a year ago interests, perhaps, only a meteorologist or climatologist. However, do you need to know the weather, or more precisely the climate, on the planet tens and hundreds of millions of years ago? It turns out, yes. Climates of the geological past are engaged in a special discipline of paleoclimatology. With the restoration of ancient climates directly or indirectly related to the solution of important practical issues. Without this, for example, it is impossible to establish the basic patterns in the location of many minerals, such as coal, salt, manganese and copper ores, and others. Obviously, the precipitation of rock salt is favored by a hot and dry climate (recall Kara-Bogaz-gol). Coals, on the contrary, accumulate in wet conditions, since the prototype of the coal seam is a peat bog. These are elementary examples.

We will not, however, go deeper into the jungle of paleoclimatology and study its physical, chemical and geological weapons. We only need to look at all those extinct plants that the story has gone and will continue to tell, in their natural environment. We met with the Carboniferous forest of Donbass, the Cordaite taiga of Siberia and the Gondwanian plants, but almost nothing was said about the climate (DeVore & Pigg, 2007).

At first glance, this is not too difficult a problem. Plants that are very sensitive to temperature, humidity, and soils, must quickly answer paleoclimatic questions. This is indeed the case when we are dealing with the latest geological sediments. Finding in the rocks a dwarf birch, mosses, marsh horsetails, sedges, we can safely say that the climate was wet and cold. Met laurel, magnolia, and platan, walnut - there are subtropics. However, unfortunately, all the listed plants were not eternal. Descending the "cut", first into the Tertiary and then the Mesozoic deposits, we find fewer and fewer modern genera, and, on the contrary, extinct plants are found. Even if we find in the ancient (say, Jurassic) sediments plants that continue surviving now, can we guarantee the permanence of their propensities to this or that climate? Paleontology teaches great caution here. A good example can be cited from the animal world. Now elephants and rhinos are inhabitants of the tropics, and only a few thousand years ago, overgrown with thick fur, they wandered in the harsh glacial landscape in Europe and Asia.

Starting to restore the Mesozoic and Paleozoic climates, it is necessary to focus not on systematic affiliation, but on the structure of plants. It should look for signs that appear in certain climatic conditions. However, this is not easy. Firstly, it is not always possible to study a fossil plant in necessary details. Secondly, science does not know why two trees growing near each other, under identical conditions, have completely different foliage. You can say even more. We do not know why the maple leaves alone, while the birch is different, why the needles sit in bunches in the pine, and at the spruce one by one. We do not know the physiological causes of a particular plant. With adaptive signs in animals, the situation is slightly simpler: the predator has claws, fast-moving fish have powerful muscles, a streamlined shape of the body, etc. The adaptive meaning of most properties of plants is still unknown.

FORTUNE-TELLING ON THE RINGS

Every student can determine how old a tree is. To do this it is necessary to count the number of annual rings on the cross-section of the trunk. The essence of the rings is simple. Within a year the tree grows at different speeds, in summer with larger, in winter with less (and even does not grow at all). In spring and summer the cells separated by cambium are large, with wide lumens; The closer to winter, the smaller the cells. These are the concentric rings of small cells and are visible on the transverse slit. The width of the gaps between the rings depends on many reasons. The arid year was given out, pests or illnesses prevailed - the rings thicken. A year passed safely - the rings are laid at a greater distance (DeVore & Pigg, 2007).

According to the annual rings on the stumps, as taught in school, you can find a road in the forest without a compass: on the north side of the tree they condense. However, if the traveler tries to find a way in this way in the tropical jungle, he will fall into a mess. There are no annual rings. Of course, not all trees and not all tropics. After all, rings (here it is better not to call them caution yearly) can be laid not only from the cold winter, as in the middle latitudes. The reason may be alternation of arid and rainy periods or just an internal rhythm in the life of a plant. The places in which the rings are found only occasionally are tropical rain forests. Such forests do not form a continuous belt, but are found by separate large spots along the equator without leaving the tropics.

In rocks, pieces of petrified wood are often found. Preservation of them may be bad, but annual rings, if they were near a tree, can almost always be discerned. There they are, which means the climate had strong seasonal fluctuations (in temperature or precipitation). When they are not there, it signals that the climate and the light regime all year round were flat, monotonous. Whether the climate was hot or relatively cool - you cannot tell by the rings.

So, one question is settled. We can allocate areas with seasonal climate changes and without them. The next question is temperature. To her approach was found on the other side. However, before - a little botanical digression and two new terms. All perennial plants with thick enough trunks, and simply speaking - trees, can be divided into two large groups. In some, and with them we mostly deal, the main part of the trunk is occupied by wood, and the bark with the core is inferior to it in volume. This type of barrel is called pycnoxyl. These, for example, are temperate trees such as pines, spruce, oaks, birches, etc. In other trees, everything is just the opposite: wood in the form of a thin ring or small areas in a continuous mass of bark and core. So palms and cicadas (sago palms) are arranged. They have a manoxylic type of trunks. Physiologists have shown that having a manoxyla trunk, the plant will not survive the winter with temperatures below zero. In other words, such trunks require a climate that is called frost-free. Speaking about the advantages of trees with a lot of wood, they usually remember the strength of their trunks, the ability to carry out a large number of vital juices. To this must be added and their ability to withstand severe winters (Leppe, et al., 2007).

Annual rings, plants with different amounts of wood, are the main trump cards of paleobotany, when they undertake to restore the climate of distant geological epochs. You can use other observations. For example, a large variety of plants, especially tree-like ferns, large leaf sizes, reproduction organs that do not sit on branches, but directly on trunks (like a coffee tree) - all this speaks rather of a tropical climate. Only in the tropics there are mangroves, that is, thickets on low-lying, periodically flooded banks (Fig. 28).

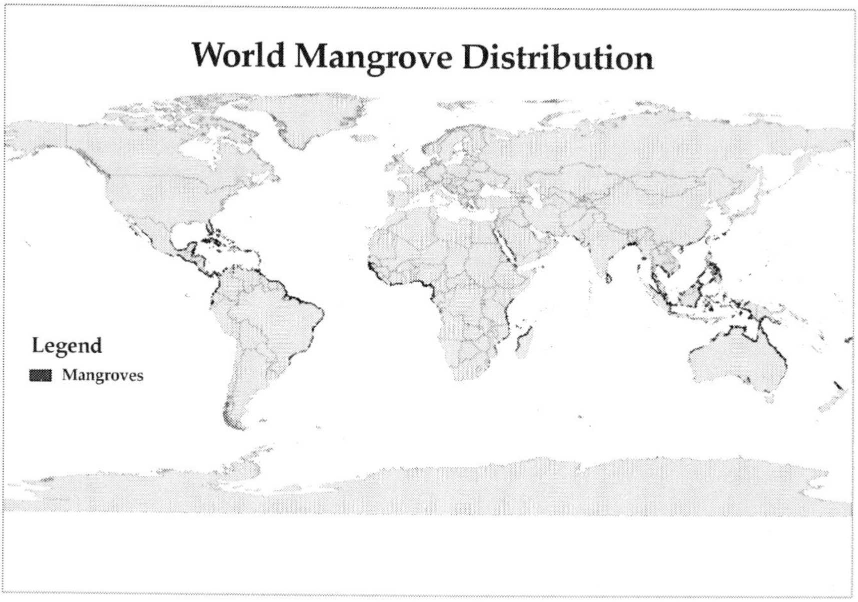

Figure 28: The spread of modern mangroves almost does not go beyond the tropics.

Of course, when judging about certain average annual temperatures, the amount of precipitation in millimeters is still an inadmissible luxury when we are dealing with Mesozoic and especially Paleozoic times. For such distant epochs, we are only entitled to delineate the tropical regions, and everything that lies beyond them is not yet better classified. All the rest for us so far are considered extratropical areas. This is not enough, of course. Is it worth it to fence for this accuracy? Yes, it is definitely worth it. In this we will see in the next chapter (for the second time it is necessary to refer to it) (Leppe, et al., 2007).

ARCTIC TROPICS

Now let us return with all the paleoclimatic weapons to the coal-mining forest of Donbass. First, there are already about 400 species of fossil plants. From the point of view of paleobotany, this is a lot. After all, we must take into account that only a few plants get to the "cemetery." It is not necessary to doubt that this is only an insignificant part of the former wealth. The story of the Carboniferous forest began with giants of lepidodendrons. Three dozen meters, and only up to the first branching of the trunk, two meters in diameter - the size is solid. Such a machine will not grow in one year. For this, the lepidodendron, despite its simple domestic economy, would have to grow several times faster than the current record holder in the growth rate, bamboo. In other words, lepidodendrons lived for many years and had thick trunks. To become necessary to us manoxylic plants, they need only have a little wood and more bark and core. All this is also evident in them. Lepidodendrons are not the only plants in Donbass with such trunks. There were many more. Some of them, cones with sporangia, grew directly on the trunks. This is also a feature of tropical plants.

Therefore, we already have three proofs of the tropical climate in the Donbass in the Carboniferous time (400 species, manoxylic plants, cones directly on the trunks). There are woods in Donbas and good preservation without annual rings, and traces of extensive mangroves. It seems everything is clear, the tropical climate, a lot of precipitation (otherwise there would be coal), minimum seasonality - in short, real tropical rain forests (DeVore & Pigg, 2007).

The same type of flora, as we have already said, occurs in North America, Western Europe, the North Caucasus, Central Asia, and North Africa (not in the whole of Africa, but only in the North). In China, Korea, and Southeast Asia, in the Carboniferous period, there were many plants that do not exist in all the places listed, but the climate here was the same, typically tropical. If the mention of tropical flora in China, Southeast Asia and North Africa does not cut the rumor, then this cannot be said about the rest of the place. Now there are at best a subtropics, and for the most part a moderate climate. However, this is not the most surprising thing.

At the end of the last century, systematic studies of the Arctic began, which continue even now on an increasing scale. Polar expeditions brought to the scientists a diverse and interesting material. However, perhaps, the most unexpected facts, which put the researchers, if not at a dead end, then in very complicated puzzles, were fossil plants harvested in the Arctic. In Greenland at 70 ° N latitude in the Cretaceous sediments are found the leaves and fruits of the bread-tree, a typical tropical plant of Southeast Asia. Tropical and subtropical genera were found in Spitsbergen, Alaska and in Greenland in tertiary sediments. True, here also found residents of temperate latitudes (chestnut, maple, etc.). For the Carboniferous time, the same picture was obtained: in Northern Greenland, in Spitsbergen, in Newfoundland, the same plants are collected as in Europe, including pieces of wood without annual rings. Of course, the number of species found here is small, but this is clearly due to the insufficient number of samples collected (Gensel, 2008).

THE GONDWANA GLACIATIONS

In the list of questions that have long been posed by science and for which it is not yet possible to give even an approximate answer, there is a question of the causes of the great glaciations. The remnants of the last glaciations, once occupying the entire north of Europe and Asia, as well as parts of the southern continents - are the ice of the Arctic and Antarctic, permafrost, tundra. In fact, it is not known whether we live at the end of the great glaciations or just in the next warming period, which for the last several million years has already been many. For the future of humankind this is an important issue, given the high speed with which glaciers come and go. Solving such a question, it is tempting to look deep into the geological history, to find similar events and with their help to understand the present situation. Unfortunately, while the reverse process is in progress: using data on the latest glaciations and modern glaciers, geologists are trying to identify and study the glacial epochs of the distant geological past. The glaciations in the life of the Earth have been repeated many times. The oldest were still in the pre-Paleozoic time, that is, over 600 million years ago. Large glaciations occurred at the end of the Paleozoic, in the second half of the Carboniferous period. This will be discussed now.

The discovery of glacial deposits of the Quaternary period was one of the most important achievements of geological thought of the last century. The final phase of geological history immediately appeared in a new light. Many of the newest deposits that geologists often dealt with turned out to be of glacial origin. All these observations were animatedly discussed in the press. It is understandable that they began to look for traces of glaciations in older sediments. Of course, immediately a little bent stick, taking for glacial boulders usual conglomerates. Perhaps, remembering about such mistakes, for a long time questioned the message of the geologist V. Blanford, made back in 1859 about the discovery of Upper Paleozoic glacial deposits in India. Some boulders in these deposits have a weight of 30 tons. Later the same deposits (they are called "Tillites") were found in Australia, Africa, South America and Antarctica.

This is how the idea of the extensive glaciations of Gondwana in the second half of the Carboniferous period entered into science. However, it is necessary to make a reservation: the time of the Gondwana glaciations is not well known. Some geologists believe that it was in the middle of the Carboniferous period, others - at the end of it, still others - refer to the glaciations already at the beginning of the Permian period. The latter point of view seems to be refuted by modern data, but to satisfy the most demanding critic, it is better to adopt a more vague formulation - "the second half of the Carboniferous period."

Although glacial epochs are not part of our task, we deliberately touched the Gondwana glaciations, since it coincided with the early stages of the development of the glossopteria flora. What does this flora say about the Gondwana climate? This question, of course, may seem superfluous when there is such evidence as glaciations. Did not the glaciers form in the tropics? However, as Commissioner Meghre said, "until the investigation is over, it is difficult to discern what is interesting and what is not." Some glaciers are not yet complete evidence of the climate, we must collect other facts (Leppe, et al., 2007).

Gondwana plants have been studied for a long time. In India, this is handled by a group of highly skilled paleobotanists and even there is a special Paleobotanical Institute (the only one in the world) (Gensel, 2008). Therefore, plants that are associated with glaciations, we know quite well. First of all, a relatively small number of species are conspicuous. The most complete list of the best-studied Indian Upper Paleozoic flora includes a little more than 160 species - and this after nearly a hundred years of intensive study. There are no plants with large spreading leaves, ferns are rare, almost no placentae, and those that have had rather thin trunks. The rocks are clogged with leaves of glossopterites. It is very important that this type of plant is found in Gordano just below the glacial deposits, above them and further along the entire section of the multi-meter coal-bearing strata. In Gondwana wood, we always see distinct annual rings.

In short, there is no need to say that the Gondwana plants grew in a tropical or even subtropical climate. It is difficult, of course, to imagine such dull landscapes in India, South Africa and South America, but what can you do! These are the facts. Surprised, we need to explain them.

THE ANCIENT MAINLAND OF ANGARIDA

In the distribution of climatic zones of the Earth, there must necessarily be symmetry. Along the equator - tropical, at the poles - the Arctic climate, and between them - something in between. Of course, there is no identical sequence of belts of different climate in both hemispheres and cannot be, but some kind of symmetry is necessary. No wonder the Earth is a rotating ball.

If so, then the question arises involuntarily, and was there not a paleozoic glaciation in the Northern Hemisphere? It must obviously be looked for somewhere outside the above-mentioned belt of the tropical flora. Certainly, the northern glaciations, simultaneous with Gondwana, we do not yet know, but this does not mean that it was not. Specific rocks formed from the melting of ice with boulders in the sea (and not on land, as in India or Africa) are known in Verkhoyansk, but they are likely to have Permian age. Perhaps this is an extra trump card in the hands of supporters of the same age as the Gondwana glaciations, but one should not rush to conclusions.

If the question of the Paleozoic glaciations of the Northern Hemisphere is still unclear, then the existence of an extratropical flora is beyond doubt. It inhabited the ancient continent of Angarida, located on the site of North Asia and at times capturing the Pechora basin. With the Angara flora (strictly speaking, it should be called "Angarid flora"), we already were acquainted with the chapter "Cordaite taiga." Solve the question of the climate in which she lived, not easy. At first, there were tree-like plauniform (though not the same as in typical tropics), which then disappeared. Then began to come across (though rarely) some tropical plants. At the beginning of the Permian period, they disappeared, and in the second half of Perm there were again quite thermophilic plants. In the Pechora Basin and Prikamye in Perm, several genera of tree-like plaques appeared, which were just as unexpected and extinct (Leppe, et al., 2007).

From the general acquaintance with the Angara flora, it seems that it is much more thermophilic than the Gondwana flora, especially in the Carboniferous. The Permian and Carboniferous flora of the Angarids are quite rich. In 1958, M. F. Neiburg counted about 600 species in them, and after that a few dozen species were described. This is slightly less than in tropical floras (only in the Donbass and only in the Carboniferous deposits found about 400 species). However, we must take into account that the Donetsk flora has been studied much more fully than the Angara flora. Yet, we cannot talk about the tropical climate on the territory of Lngaridy, when there was a cordaite taiga on it.

First, such a variety of species is characteristic only of the Kuzbass, the Pechora Basin and the Far East, i.e. for those areas that are located close to the tropical belt. Already in the Tunguska basin the variety of plants was much smaller. In addition, all Angara wood has good annual rings. Apparently, there were almost no treelike ferns here. Therefore, it is possible to speak of the climate of the Kordaite taiga as a vetropic one. Apparently, at the beginning of the Carboniferous period, when there were no Cordaites yet, in the south of the Angarida (in the Kuznetsk and Minusinsk basins, Tuva and other places) there were tropics, since the main plants found here in the Lower Carboniferous sediments are fairly large (up to half a meter In diameter) tree-like placentae (Leppe, et al., 2007).

WATER, WATER!

In some deserts, the amount of precipitation is so small that it cannot be measured. For example, in the Libyan Desert, rain occurs once every four to five years. Yet the deserts are populated with plants. To combat dryness, they develop unusually deep roots penetrating up to 15 m deep into the soil. In other plants, the roots, on the contrary, spread across the area, stretching for a distance of 15-20 m. Drought resistance is one of the most interesting problems of plant physiology. Actually, the whole evolution of plants from the moment they reached the dry land went under the sign of the struggle for water. Here you can find interesting analogies with vertebrates, who also once left the sea. The most primitive terrestrial plants (and among the animals - amphibians) need their water for their reproduction. Frogs have to throw caviar somewhere, and spermatozoa even of such terrestrial plants as ferns and plaunas can reach the female egg only on the water (true, for this they have enough hot water film). Reptiles, birds and mammals need water only for drinking (if they do not live in water). Those who eat juicy vegetation may not drink at all. The most complex plants (gymnosperms and angiosperms) also do not need water for reproduction. In short, the organic world is trying to become more and more independent of water.

Unfortunately, we do not know how the higher plants developed such complex adaptations that allow them to reproduce in almost completely dry places. The fact is that seeds, i.e. the main adaptation for such reproduction, appear in geological history somehow very suddenly, without a gradual transition. For now, only hypotheses are being constructed about their origin (Gensel, 2008).

However, this is not the point now. On the previous pages, we analyzed the climatic conditions of past eras, talked about tropical and extra-tropical areas, paying attention to the temperature and seasonality of the climate, almost nothing was said about the humidity. Meanwhile, the moisture of the environment should give the most complete information about plants. If the plants responded primarily to the moisture of the environment, then in the fossil flora you can find the appropriate patterns. Indeed, we know not only individual plants of dry climates (for example, in the tertiary deposits of Utah in the US, a cactus was found), but also the whole flora. Dryness has left its imprint on it in everything: leaves are small and thick, many conifers, ferns and horsetails are rare. It is precisely such sets of plants that we find in the Upper Permian sediments of Western Europe, where for a hundred years of careful study only a dozen or three species have been established. Together with such "oppressed", as it is often said, flora in the geological section, there are gypsum, anhydrite, rock salt, dolomites - in a word, rocks formed in the conditions of a hot, arid climate. For a long time, geologists have been misled by rocks that are painted in bright red and brown colors. They considered that this was nothing but petrified deserts. Later found in these "desert" rocks of the shell of aquatic animals. Nevertheless, a large number of red-colored rocks really proved to be evidence of the arid climate (Leppe, et al., 2007). Now the dry-loving flora is established in the sediments of all periods, beginning with the Carboniferous, and more or less the outlines of their distribution. However, the question is not exhausted.

PLANTS-PERJURERS

It seems that everything is logical. Plants depend on the humidity of the climate. In arid conditions, we must meet and dry-loving species. Naturally, the opposite conclusion: if we find dry rock plants in the rocks, then the climate was droughty, and if the moisture-loving plants get into our collections, then the climate was wet. Such reasoning looks like axioms. Why break through the open door? However, it is believed that there are no rules without exceptions. If you talk about the geological past, you often do not know where the rule is, and where the exception is. Some kind of insignificant amendment or one exception suddenly casts doubt on a completely logical theory, and in the end it is completely destroyed.

Approximately so it turns out and with the just stated axioms. They do not take into account one thing: what inhabited the territory of the region millions of years ago, and what fell on the paleobotanical table are completely different things. The inhabitants of Central Asia are familiar with such a picture. Spring comes, and the severe desert begins to turn green. Then the sun is hot, and the time comes for the summer dull landscape. Everything around dries, there is only a scanty desert vegetation. Only oases and river valleys are green. Is this a lot, if you count as a percentage of the total area? In addition, what will the paleobotany of the future take? We have already said that a plant cannot become a fossil, but after falling into the water. There are, of course, exceptions, but this is a rarity. For this, for example, there must be a volcanic eruption, so that ashes fall indiscriminately. Sometimes trunks come up to us, covered with dunes, but there is only rubbish left from the leaves. Therefore, it turns out that in arid regions there are chances to become paleobotanical documents are not dry-loving plants, namely those that grow along water bodies. Of the remaining plants, only pollen will be preserved.

Here you can, of course, argue that this is all also just a reasoning. We need examples, and they are. In the second half of the Permian period, the eastern half of the European part of the Union was closed by a large bay of the Arctic Sea. Then the sea left from here gradually, leaving here and there salty lagoons. The climate was hot and dry, and therefore sediments began to be deposited on a large scale in the bottom of the lagoons: dolomites, rock salt, gypsum, etc. However, among the typical rocks of the arid climate, there are interlayers with a strange set of fossil plants. These are horsetails, mosses, ferns, plumaceous - in a word, very hygrophilous plants. This is not an isolated case.

Geologists who have a collection of fossil plants often ask the paleobotanist not only about the age of the rocks, but also about the climate in which they found plants lived. It is important for the geologist to know the situation in which the rocks were deposited. Answering such questions, paleobotany has to be very cautious. Of course, if we find a large number of mosses and horsetails in the rocks, we can almost certainly say that these plants required a lot of moisture for themselves. However, if among the fossil plants there are no such characteristic indicators of the humidity of the situation, what to do then? As in the case of temperatures, we must look not at the systematic belonging of plants, but on their shape, structure, and microstructure (DeVore & Pigg, 2007). So paleobotanists did, using observations on modern plants. However, it turned out to be an incident.

At first everything went smoothly. Having at hand typical examples (cactus, saxaul, feather grass, etc.), botanists created an abstract image of a dry-loving plant. It, they say, has small leathery leaves, a thick cuticle, hidden in holes or grooves of the stomata. In plants such as cactus or euphorbia, leaves are few, fleshy, shoots are also fleshy, many thorns, etc. Despite numerous exceptions, it is these types of dry-loving plants that have settled in the consciousness of paleobotanists as the only authorized representatives of all plants adapted to the defect Of moisture. A hasty reverse conclusion was made: if the fossil plant found contains all of the above signs, then it grew with a sharp lack of moisture. Over the past 50-odd years in the press, there have been articles calling for caution in such conclusions. Cogent facts are also presented. With some of them, we now are acquainted.

WHEN THE LIGHT BECOMES AN ENEMY

Berries of cranberries are known to everyone, but its leaves are not properly scrutinized, although they come across even in factory jars with jam. The leaves are very small, similar to cowberry (both plants belong to the same genus of the heather family), but smaller and stiffer. By all the signs that we recently listed, cranberries are a typically dry-bred plant (xerophyte), despite its marshy lifestyle. There are many such plants in the marshes, and this should have alerted paleobotanists. Hence, plants with "dry-loving" appearance can be found among the inhabitants of the marshes of past geological epochs. Finding such leaves, you can come to the conclusion of a dry climate - and get into a mess.

This happened when they began to study the plants of the European bogs of the Carboniferous period. Paleobotany found signs of drought resistance on plants, and geologists saw in the rocks all indicators of the wet climate. To satisfy both sides, we had to make a kind of compromise. The climate of the Donbass in the Carboniferous period in one of the recently published books says that here, according to paleobotanical data, the climate was, allegedly, hot and dry. This is indicated by xeromorphic signs of some plants. Intensive coal formation in this era is, on the contrary, evidence of a damp climate ... Most likely, periods of dry and humid climate alternated. Thus, the contradiction was not so much resolved as it was covered with unjustifiably complex constructions. After all, the alternation of wet and dry climate (climate, not weather) is a very unusual situation.

The authors of this book can be understood. Without data on the physiology of modern bog plants, they naturally could not imagine such nonsense: the plant armed against dryness, living on a swamp surrounded by water. We will not blame them for the reason that the botanists themselves, including specialists in the physiology of plants, did not quite understand the reasons for this paradox. There are, however, a few fairly convincing assumptions (DeVore & Pigg, 2007).

At the end of the last century, the German botanist A. F. Shimper came forward with his theory of physiological dryness. He and subsequent researchers have shown that the plant cannot always take advantage of water, which seems to be in abundance. Such physiologically dry are, for example, solonchaks. With amateurs of saline places paleobotany to meet it was necessary. This, for example, Triassic plant pleuro-meia - a distant relative of lepidodendron. By the way, the history of her first find is unique. In the middle of the last century, a stone fell from the cathedral in Magdeburg and, naturally, it split, and in it a piece of pleura was found to be well preserved. According to the nature of the rock and the structure of the plant itself, the paleobotanists came to the conclusion that the pleura formed a continuous thicket in shallow saline basins. This is only one type of physiologically dry habitat.

Sometimes it is believed that in the marshes water is not suitable for all plants because of the chemical composition (there are many humic and other acids in it). In addition, marshes freeze in winter and evergreen bushes, like cranberries, are deprived of soil moisture. In order not to perish, they must protect the water that is in their tissues, and for this purpose develop devices that reduce evaporation. Xerophytism of marsh plants seemed to have received an explanation, and it was this view that was adopted by paleobotanists studying similar traits in plants of fossil bogs. However, unfortunately, and in these arguments flaws were discovered.

Some botanists (especially worked hard in this direction MV Senyaninov-Korchagin) found that marsh people almost completely lose their drought-resistant "uniforms" if they are grown in the same physiologically "dry" water, but in the shade. The conclusion from these observations was quite unexpected, although natural: plants try to protect themselves from light, from the source of their life. Subsequent studies have shown the cause. It turned out that with an excess of light and a simultaneous lack of nitrogenous substances, clumping takes place, and then the death of chlorophyll grains, without which the plant cannot live. On the swamp with its low-growing vegetation, light is more than enough, but there is a great shortage of nitrogenous substances in the soil. To fill this deficit, plants cannot (some plants in the pursuit of nitrogen learned to catch insects), and they can only hide from the light. Small leathery leaves, pubescence and thick cuticles are their effective protective measure (Alves, et al., 2005).

It turned out, of course, very unsuccessfully for paleobotanists (it is necessary to take into account also light), but it is successful for plants. With the acquisition of these features, they achieve two purposes: in the case of temporary drying or freezing of the swamp, plants do not suffer from a lack of water, and in the case of excess light, they do not die from the adhesion of chlorophyll grains. It is impossible, of course, to completely dismiss the physiological dryness in its original understanding. The water in the marshes is really not very good, and one thing you cannot explain with one light. Some signs of fossil (and modern) bog plants, for example the location of stomata in pits or grooves, are difficult to relate to the reaction to excess light. Therefore, those leaves of the Cordayites, in which the stomata are hidden in the grooves and whose study was the most difficult, seem to indicate rather a certain physiological dryness of the marshes in the Corday taiga.

All that has been said shows once again how closely the branches of science are closely connected and how far it is sometimes necessary to seek the solution of the questions posed by nature. We have already said that the identification of regularities in the distribution of certain minerals requires knowledge of the climatic conditions. For this, one must turn to fossil plants. Hence the logic of research leads to the physiology of nutrition of modern plants. Who would have thought that when a botanist closes the bush of the growing wild rosemary in the swamp and then measures the thickness of the cuticle on its leaves, it helps to reveal the underground wealth? It is appropriate to quote the words of the famous French biologist Lanark: "The man is subtle in some object, a vast, even, but closed, can his opinions on the subject to bring us evidence and his experience, and his knowledge of all the details of the area that is the subject of his observation and study. However, if it is little diversified range of their ideas and knowledge, if he remained alien to the majority of the ideas and knowledge that he could get from another source, then it is actually a loan t on the stairs of various stages of intellectual development of a moderate height level ... and even he cannot afford to understand themselves or to establish the true philosophy of science developed by them. "

DID PLANTS HAVE A GOLDEN AGE?

You can answer firmly: yes, and maybe even several times over. However, we must agree on terminology. The concept of the "golden age" is somewhat vague. We define it by contradiction (Lieberman, 1987). The golden age can be considered as the time when from the north and south of the glaciations the vegetation zones were not pressed to the equator, when there were no giant arid zones, when the rainforests were not squeezed by rigid frames near the equatorial strip. Of course, prosperity for some plants means death for others, but we will proceed from the interests of the majority. According to the number of species, tropical plants constitute and have always been the overwhelming majority in the plant world.

Having accepted this definition of the golden age, let us see when the plants were most fortunate. The first time it was, it seems, in the first third of the Carboniferous period. At that time tree-like Plauen and other witnesses of typically tropical conditions spread throughout the world. A frost-free but slightly dry climate was then in the Angarid, i.e., where the cordaite taiga later grew. From the second third of the Carboniferous period, a sharp differentiation of climates began, which grew more and more. On the Gondwana continents, a great glaciations began, a distinctly diverse flora inhabited distinct climatic zones. In the Permian period, an extensive belt of arid climate formed, which lasted also in the first half of the Triassic period (the first in the Mesozoic era) (Griffen-Foley, 2002).

Great events occurred in the life of plants in the second half of the Triassic period. With a monotonous tropical climate, they have opened wide paths for resettlement. Almost all over the world we find a similar set of genera and even species, whether it is Northern Greenland, Central Europe, the Urals, China, North America, the Far East or South-East Asia. Some peculiarity still exists in the flora of the Gondwana continents. However, in the Jurassic period, the southern hemisphere is gradually capturing the floristic balance. Of course, one should not think that the flora of all the places listed completely repeats each other. The difference in the generic and species composition, of course, is, but can it be compared with those pronounced differences that were in the Permian period or which are observed now. For the end of the Triassic period, it has not yet been possible to establish climatic zonality for fossil plants (this conclusion was recently arrived by the palaeobotanist IA Dobruskin). In the Jurassic period and in the first half of the Cretaceous period, climatic zoning is noted, but very heat-loving (maybe even tropical) plants are taken far into the depths of the extratropical areas. There are plants that are found almost all over the world.

Such very favorable climatic conditions of the planet continued for a long time, about one hundred million years. Even then, already in the Tertiary, in Western Europe there were almost real tropics. Not far from London, in our Ukraine and in many other places find the remains of palm trees, ferns, various evergreens, whose immediate relatives are now found only in the tropical and subtropical jungle. All this prosperity did not become, when the last glaciations began, which drove away all lovers of heat far to the south. The golden age is over. There were vast spaces with forest-tundra and tundra, which before, perhaps, did not form a continuous belt. Arctic plants before, perhaps, lived only in the highlands or the Far North.

We are accustomed to approach all events of past geological epochs with the yardstick of the contemporary face of the Earth. However, the further we go deeper into the history of our planet, the more clearly we see that our geological period is the exception rather than the rule in the climatic sense. Such a sharp climatic zonation of floras, which can be seen on Earth now, may have been in the history of plants only once, namely, at the end of the Paleozoic era. This is one of the most interesting conclusions that geological science has come to lately with considerable assistance from paleobotany.

Unfortunately, we know little of the early history of the plant world. Specialists on Devonian plants often say that for those times it is not possible to identify any floral areas, to outline patterns in the geographical distribution of plants. Perhaps this is only an apparent monotony due to the lack of our knowledge. However, the fact that the better-studied Early Carboniferous flora fits only into two or three regions, which have much in common with one another, indicates that the climatic zonality was not too sharp in the Devonian. If so, we get an interesting general sequence in the development of the terrestrial plant world. At first there was a poorly expressed zonality, then it manifested itself brighter and stronger, there was an ice age, then the desiccation of large spaces. Then there is a sharp decline in the differentiation, again a relatively monotonous set of plants settles over the entire Earth (Griffen-Foley, 2002). Then climatic differences of zones again grow, at the culmination point the giant glaciations breaks out, and. . . Further we do not know what will happen. To derive regularity and predict the future, there are few two cycles, one of which we do not know the beginning, and the other - the end. However, to reject the possibility of returning the next "golden age" is impossible. Unfortunately, in any case, this will happen no sooner than a few million years.

REFERENCES

1. Griffen-Foley, B. (2002). The Fairfax, Murdoch and Packer Dynasties in Twentieth-century Australia. *Media History*, *8*(1), 89–102. https://doi.org/10.1080/13688800220134536
2. Lieberman, V. B. (1987). Reinterpreting Burmese History. *Comparative Studies in Society and History*, *29*(1), 162. https://doi.org/10.1017/S0010417500014390
3. Alves, L. S. D. R., Guerra-Sommer, M., & Dutra, T. L. (2005). Paleobotany and paleoclimatology. *Topics in Geobiology*, *23*, 179–202.
4. DeVore, M. L., & Pigg, K. B. (2007). A brief review of the fossil history of the family Rosaceae with a focus on the Eocene Okanogan Highlands of eastern Washington State, USA, and British Columbia, Canada. In *Plant Systematics and Evolution* (Vol. 266, pp. 45–57). https://doi.org/10.1007/s00606-007-0540-3
5. Leppe, M., Fernandoy, F., Michea, W., Munoz, C., Palma-Heldt, S., Raymond, C. E., & International Symposium on Antarctic Earth, S. (2007). Paleobotany of Livingston Island; the first report of a Cretaceous fossil flora from Hannah Point. *Open-File Report - U. S. Geological Survey*, Short Research Paper 081. https://doi.org/10.3133/of2007-1047.srp081
6. Gensel, P. G. (2008). *The Emerald Planet: How Plants Changed Earth's History. The Quarterly Review of Biology* (Vol. 83). https://doi.org/10.1086/586952

7 FOSSIL PLANTS ABOUT DRIFTING CONTINENTS

Science has many problems. Some are more attributed to science itself, while others are often underestimated. How much paper was used to discuss the "snowman" and "Loch Ness monster", to which God knows what level scientific significance they showed. Meanwhile, the discovery of live coelacanths, the real fish kistepers, that is, the ancestral form of terrestrial vertebrates, for all the sensationality of this catch, pockmarked on the pages of popular publications, left no noticeable trace in science, did not give a stunning material for paleontology. It turned out that paleontologists on the bones had figured out in kistepery fish quite well. Latimerium only confirmed the existing views (Fisher, 2011). At the same time, the greatest riddles of nature (some we have already talked about, others we will meet again) have remained poor orphans in popularizers and will not come into fashion in any way. This has its pluses - the riddle cannot be dragged and vulgarized, it makes it less attractive. However, this and the minuses - even among researchers these puzzles are often forgotten.

In this sense, the hypothesis of continental drift was lucky. Originating in scientific publications, it went to the pages of the semipopular (albeit strictly scientific) book by Alfred Wegener and quickly attracted everyone's attention. Unlike many other hypotheses, it equally excites specialists and reading public, and geologists have sharply opposite feelings. Some have enthusiasm and enthusiasm, others have irritation, unwillingness to deal with it. This hypothesis remembers the labels of "idealism" and "catastrophism," it was reproached for "leading scientific thought into a world of ghosts and self-deception." Yet it was extraordinarily tenacious.

For the first time the possible movement of the continents was seriously started in the 18th century. (That is really true that the new is a well-forgotten old). In § 163 of his brilliant book On the Layers of the Earth Lomonosov writes about finds in the northern regions of remnants of plants and animals characteristic of warm countries. These are "sea crabs," that is, mollusks, "elephant teeth" (mammoths), "traces of Indian herbs." Some of these observations are explained by the testers of nature in two ways, others think more of the main globe of transformation, the great part of which has been moved from place to place by the extreme violence of the inner underground action, others attribute to the insensitive inclination of the whole earthly globe that in many centuries changes the distance of the ecliptic from the pole. " About an hundred years later the anonymous author wrote in his book published in Livni: "In the outline of the continents of the globe, the following is outlined: the western shores of Europe and Africa are almost parallel to the eastern shores of America (Fisher, 2011). All this can hardly be a mere coincidence." True, the pioneers in the theory of displacement were, apparently, still the Abbot of Plas, according to which, referring to 1668, "before the flood America was not at all separated from the Old World." Thus, this hypothesis is already three centuries old, and everyone is arguing and arguing about it. More and more geologists, geophysicists, paleontologists, botanists, oceanologists become its adherents.

Paleobotany never stood apart from the discussion, and this is understandable. On floating continents there were always passengers of plants, which could not help reacting to movements along a degree grid, could not be transplanted from the mainland to the mainland when they moored to each other. Similarly, the thousand-kilometer oceans between the continents had to affect the patterns in the distribution of plants. If we collect all the paleobotanists in one room and ask them to vote on the question of continental drift, Wegener's supporters will undoubtedly constitute the overwhelming majority. This is understandable, since it is very difficult to reconcile the facts from the geography of fossil plants with the permanence of the present situation of the continents. With them we already have to know a little and now we will return to them again, but from a slightly different angle. Take as an initial point the end of the Carboniferous period and the very beginning of Perm (Rothwell & Nixon, 2006).

PARADOX NUMBER 1

For a selected period of time, we can build a fairly accurate picture of the distribution of the flora of different climates. In the middle there was a strip of tropics: almost all of North America, Greenland, North Africa, Western Europe, most of the European Union territory, Asia Minor, Central Asia, China, Korea and Southeast Asia. Draw the boundaries more precisely, relying on the location of fossil plants of this age and look at the resulting map (Figure 26).

The picture as a whole is familiar - about the same as now. In the middle - the tropics, at the edges - extratropical flora. Surprise is caused only by the outlines of these zones. In our view, the tropical zone is a more or less regular band along the equator (Figure 29). On a modern climate map, this regularity cannot be disturbed even by the very uneven location of continents and oceans (in the Northern Hemisphere the area of the continents is much larger than in the Southern Hemisphere) (Rothwell & Nixon, 2006).

Figure 29: The spread of modern tropical rain forests with trees without growth rings (1) and the spread of tropical flora in general.

Even a close similarity of this correct picture, we do not see on our Permian-Carboniferous map. On the longitude of America and Western Europe, the width of the fossil tropical zone is 6-8 thousand km. On the longitude of the Pamirs, the zone narrows to a thousand kilometers. However, this is not the main thing, almost the entire tropical zone is located in the Northern Hemisphere! Only in Indonesia and, perhaps, in South America, it goes a little beyond the equator. However, the Southern Hemisphere is almost completely occupied by a single extratropical, rather cold-loving glossopterian flora. By itself, this fact causes distrust of fixation, that is, to the theory of the stability of continents. Palaeobotanists cannot imagine that the same plants (not individual genera and species, but rather rich flora) could safely exist 450 km from the South Pole and 30 ° north of the equator in Kashmir and at the same time at an even greater length In the perpendicular direction (Archangelsky, 1996).

Of course, "can not imagine" is not yet a very weighty argument, since the fixers can not imagine the movement of the continents in the same way. The most orthodox of them because of this can no longer convince any arguments. Still here we are talking about representations of different orders. When paleobotanists talk about the impossibility of such a spread of plants, they rely on well-known patterns of their geographical distribution. We already know enough about fossil plants to categorically assert that they had in principle the same set of tissues and cells as modern plants. We can give a head to the cut-off that the terrestrial plants from the Paleozoic to the present have preserved their needs in the external environment, have preserved the norm of reaction to this environment. In the first chapter, we said that plants have typed their "lexicon", which they speak to the outside world, back in Devonian times. Therefore, paleobotanists have no reason to confer vegetation on the plants of the geological past with magical features that allowed them to ignore geography, settle as it is convenient for the supporters of the constancy of continents (Archangelsky, 1996).

When the supporters of fixism say that they can not imagine something, we are talking about very different matters. They can not imagine the processes that have been going on for millions of years on the inaccessible depths of our planet. There is often, in fact, nothing to represent here, so little we know what is being done and done at depths of hundreds of kilometers. No physical reasoning can be taken into account here until it is clearly demonstrated that the physical laws that were taken as a basis are valid for hundreds of millions of years. After all, for such a long period of time there are processes that are also quite difficult to imagine, approaching them with usual time measures. Could we imagine, if we did not see this in nature that the layers of hard rocks, like flint, can bend into whimsical folds, as if they are fashioned from plasticine? However, back to our subject.

From the south, the tropical zone encompasses the half-ring Angarida with its extratropical flora. The area of Angarida is 10-15 times smaller than Gondwana (if the oceans dry between the southern continents). Again this is some kind of nonsense. It is also interesting that the Angara flora, apparently, was more thermophilic than the Gondwana flora. Probably, this is because there were no glaciers here, and if they were, they were small and little influential.

In modern climatic zones asymmetry is also observed. In the southern hemisphere there is almost no temperate zone, here the tropics and subtropics are very sharply replaced by tundra. However, the current asymmetry almost does not concern the tropics themselves, as can be clearly seen on maps of the distribution of the tropical climate in general and individual characteristic plants, for example palms. Present tropical rain forests with their timber without annual rings are also located the right strip along the equator. On our Permian-Carboniferous map everything is not so, and this is the first paleobotanical paradox.

PARADOX NUMBER 2

The influence of warm and cold currents on the climate is well known. The best example is a comparison of Southern Greenland and Scandinavia, located on the same latitude. The difference in their climate, due to their origin of the Gulf Stream, does not need explanations. Therefore, it is quite understandable that it was precisely to the turnings that the researchers addressed in search of an explanation of the paradox No. 1. At that, they gave both unfounded assurances that currents can all explain, so attempts were made to seriously paleoclimatic analysis. Add to the circulation of the atmosphere, select the mountain ranges along their way, and we will get a rather plausible, from the point of view of fixation, picture, which was drawn, for example, by the prominent English climatologist G. Brooks.

However, with the currents, the circulation of the atmosphere and the mountain ranges, things are not as simple as they seem at first glance. For example, in Europe the northern boundary of the tundra under the influence of the Gulf Stream is located only 10 ° north of Siberia. This action is the most powerful modern heat current (Herrera, et al., 2010).

Here the skeptic will notice: "Well, what is the Gulf Stream? There used to be currents! And now, after all, times are different, and the scale is not the same." It is difficult to object here, and therefore simply turn to facts. We will not, at our desire, set up here and there ridges and draw logically possible variants of circulation of air and water masses. Let us look at the documents, insofar as the currents, ridges and so on actually affected the location of the climatic zones.

In recent years paleobotanists have made many efforts to understand the main regularities of the geographical location of fossil plants. The famous Moscow paleobotanist VA Vakhrameev built a series of maps for the Jurassic and Cretaceous periods, indicating the main paleofloristic boundaries. I managed to build similar maps for different parts of the Carboniferous and Permian periods. On all these maps we trace only one, the most justified border dividing the tropical and extratropical flora of the Northern Hemisphere in Eurasia. In addition, it turns out that from the beginning of the Carboniferous period and almost to the end of the Mesozoic, this border changed its position little (Herrera, et al., 2010).

For these quarter of a billion years, the face of the Earth has undergone tremendous transformations: continents and oceans have changed their contours many times, sea waters flooded most of Eurasia, and then left. The seas were replaced by deserts, the earth's crust crumpled into folds. There was no Urals, then it appeared and became an impressive mountainous country, and then again almost equaled with the earth. Mountains and mountains in Europe, Central Asia, the Far East, etc., appeared and disappeared. Mountain chains and seas that cut Eurasia made it difficult to exchange plants between different regions. During this time, whole dynasties in the plant and animal world of the planet blossomed and decayed. In addition, despite all this, the border we chose was only lazily swaying slightly to the north, then slightly to the south, but did not change the orientation and never went beyond the 1.5-2 thousand km strip (Figure 30). This is paradox No. 2, the paleofloristic boundaries of higher rank depend little on changes in the face of the Earth (Carcaillet, 2007). A more general reason - solar radiation - plays a decisive role here, Well, if so, how to reconcile the paradoxes No. 1 and No. 2 from the point of view of fixation? Why did the Sun pass in the Paleozoic with its warmth the southern half of the Globe with its cold-loving flora? It would be always good, but it is not like that!

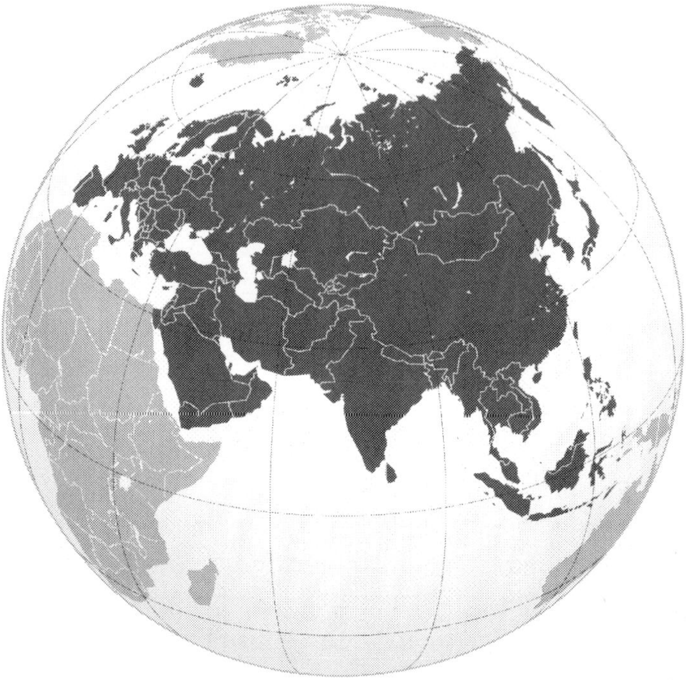

Figure 30: Northern Boundary Of The Tropical Zone Within Eurasia.

The southern equivalent of the border that we traced for almost 250 million years is not so constant. This boundary, during the geological history, stubbornly crawled south. Take, for example, a better-studied India. In the Paleozoic, it is covered with extratropical flora and, by the way, is cut off from the northern flora (more on this in more detail below). Butc in the middle and especially at the end of the Mesozoic (60-150 million years ago), India already clearly lies in the tropics. It is inconsiderate for the fixists: the northern boundary of the tropics is almost not moving, and the southern one is all leaving and going down the meridian. Both borders are directly dependent on the Sun, and for some reason it was obviously not the same for them. However, this does not stop the testimony of fossil plants.

GREAT PALEOBOTANICAL WALL

Paleobotanists and botanists, not inclined to support Wegener's ideas, have long put forward the "bridge theory." They believe that the similarity of flora on both sides of the ocean is understandable without the theory of continental drift. Plants can easily migrate through archipelagoes or simply along the waves of the ocean. The wind and migratory birds carry seeds and spores with them. These are all good arguments, supported by specific observations. However, it is not enough to reveal the factor, it is also necessary to evaluate its impact to the full. Without historical analysis, this is indispensable. We need the whole sequence of events, not just the final result. If the exchange of flora did not encounter difficulties, we will certainly see this on fossil material (Carcaillet, 2007). For example, let us take Eurasia, where the ratio of the flora of different regions is best studied. Let us not forget about other territories. So, let us repeat once again the situation that has developed in the tropical world of Eurasia in the Paleozoic. There is a tropical belt of uneven width, to the south of it - Gondwana, to the north - Angarida. Both with extratropical flora, in which little, if not at all, tropical plants. The latter is not surprising, since plants can not freely change climatic conditions (Carcaillet, 2007).

However, that is interesting: between the purely Gondwana and purely tropical flora, there is no mixture even in the border areas. In Northeast India (Assam, Darjeeling, Sikkim) - a typical glossopterin flora. Passed through the Himalayas, and in Southwest China there is also a typical tropical flora. Even more interesting is the picture in the region of Kashmir and Central Asia. In the region of Srinagar (almost the most north of India), numerous glossopteris and everything found in the Goydvan flora were encountered. In addition, less than 1000 km to the north we already meet the northern extratropical (Angara) flora. It is here in the north of India near the junction of the flora of different types on a small patch, I would like to see a meeting of the Gondwana plants with guests from Central Asia, Kazakhstan, Siberia and China. Although such an encounter was to be facilitated by the archipelago of the islands located here, there is no trace of the fact that it took place. It creates a distinct impression that India at that time was fenced off from the rest of Asia by a paleobotanical "great Chinese wall." Only the most courageous "travelers" from the world of plants penetrated through it, they did not break any short-term isolation, but independent development of the original Gondwanian flora for at least fifty million years (Carcaillet, 2007). The fact of the independence of the Gondwana flora from its northern Angara cousin was not always obvious and was fully realized only recently. A small historical digression will be appropriate here.

A person familiar only with consumer electronics, seeing an oscilloscope, probably in the soul will call it "strange TV." It so happened that to identify unfamiliar objects, we first look for a name from the available assortment of words. No wonder that the French potatoes have an earth apple. For the same reason, there are marsupial wolf and sea urchin, although these are not wolves not a hedgehog. It also happens approximately when a collection of plants completely unknown to it gets to the paleobotanical. Caution and prudence do not allow us to immediately distinguish new genera, and he tries first to find at least a generic name among already known plants, "Make" a new species is not so scary.

When, at the beginning of this century, fingerprints of a fern collected in the Kuzbas fell into the hands of the Russian paleobotanist I. Petunkpikov, he immediately saw their similarity to the Indian species of the genus neuropteria. Therefore, he called them. Later, Indian leaves were isolated in a special genus gondwandium. Here the Kuznetsk leaves crossed, though in a special kind. Until recently, there was no doubt that the Kuznetsk and Gondwanian leaves were close relatives. This was explained by the fact that none of the paleobotanists had the opportunity to personally compare them on the samples (Mendoza, et al., 2012).

In 1966 I visited India and in the Indian Museum in Calcutta I became convinced that only Indian leaves bear this name by right. In Siberian species, each leaflet has a clear middle vein, which should not be found in plants that claim to be called gonvanidium (Figure 31). There were other differences. So one link linking the Angara and Gondwanian floras burst.

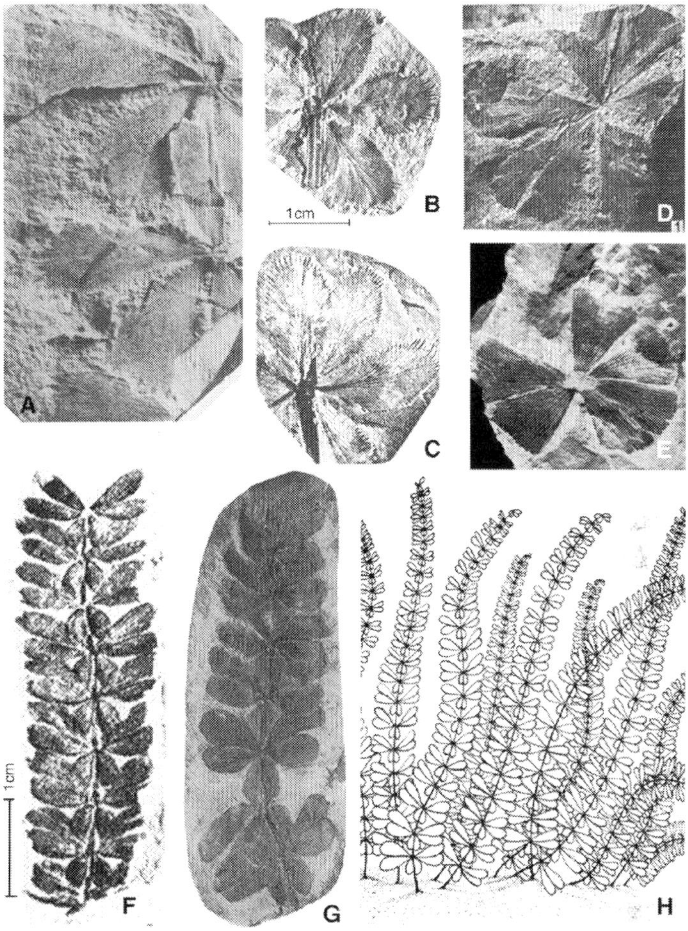

Figure 31: Leaves From Siberia And India.

For the sake of justice, it should be noted that Petunnikov did not accidentally turn to objects for comparison to India. At that time, so. At the beginning of this century, the paleontological world was strongly influenced by the findings of VP Amalitsky. At the end of the last century, he discovered on the Lesser Northern Dvina Pareiasaurus and other reptiles, which previously were known only in South Africa, i.e., in the bowels of Gondwana. Together with the Pareiasaurs, plant prints were also found which, both Amalician and French paleobotanist R. Zeyller, and our paleobotanist, MD Zalesskii, were classified as genus glossopteris. Only much later in 1933 MD Zalessky singled out them as a separate kind of pursongia, but still did not doubt their proximity to glossopteris. With them came a serious confusion, which was revealed only recently, when the microstructure of purpoids was studied and it was discovered that they had nothing to do with glossopteris. At the beginning of this century, this was not known, and there was no doubt that the Gondwana plants and animals lived in the north of the European part of Russia. In addition, the English paleobotanist EA N. Arber depicted everything on the corresponding map. Under the impression of these views I. Petunnikov, naturally, saw in the Siberian gondwandiums not only an external similarity, but also a close relationship with the Gondwana prototype (Mendoza, et al., 2012).

The issuc with the Cordayites was much more difficult. We have already said that they are a very difficult not to palaeobotany. However, Indian and Angara leaves are especially unlucky. The story began with annoying mistakes and misunderstandings. In 1879, the Czech paleobotanist O. Feistmantel, the founder of a systematic study of the Paleozoic flora of India, described the new genus of the Gondwana cordays and named it Noeggerathiopsis (neggerathiopsy). On one of the samples, several broken leaves lay parallel to each other, from which Feistmantel concluded that even during the life the leaves sat in two rows on either side of the rod, that is, they were located in the cirro. Hence the name of Noeggerathiopsis itself, since the plant neggerathia, which Feistmantel considered a close relative of the kind established by them, the leaves sit just like that. In this Feistmantel saw the main difference of the Gondwana Noeggerathiopsis from the real Cordayites of Europe. However, this was a mistake that quickly became clear. Under the existing rules, the name Noeggerathiopsis had to be abolished, which was not done in time, and then they got used to it and still many paleobotanists use it. Moreover, according to a pure misunderstanding, the Angara Cordayites were also called and still are often called, although there was no reason for it. It turned out to be like a hard-working old gossip. Such confusion is not only a matter of the purity of the paleobotanical language. After all, for every Latin name should stand a certain appearance of the plant. Otherwise, serious misunderstandings can not be avoided. Therefore, it happened with the Cordayites of Gondwana and Angarids. They figured under the same, and moreover also the illegal name, and confused paleobotanists. It is not always possible to unearth the source of the error, and not everyone likes to carry out such an investigation. When the building is built, it is very difficult, and I do not want to replace a brick in the foundation (Carcaillet, 2007).

This bunch of misunderstandings and mistakes supplemented the horsetails. Talking about the Cordaite taiga and glossopterin flora, we casually touched horsetails with cupped-shaped, fused at the base of the leaves. They will be discussed in more detail in Chapter IX. Both Gondwana and Angara species were boldly related to the same genus, and recently it turned out that they can not be attributed even to one family.

Therefore, there was a steady belief about the significant affinity of the Angara and Gondwanian floras. With this belief we have to say goodbye. Do not just hurry up and conclude that they really have nothing in common. Some plants, like Athanasius Nikitin, bravely moved from one area to another, but were always lost on an alien background. The fact that such immigrants play an insignificant role was also not taken into account when comparing both extratropical floras. After all, the list of plants rarely indicates how many instances of each species were found - one or several thousand. I happened in India to visit coal deposits and see clusters of many thousands of leaves glossopteris. In Siberia, this genus also occurs, but it is a rarity. Only a few fingerprints are known to the Palaeobotanists (Fig. 32).

Figure 32: Leaves glossopteris from the Upper Permian sediments of Siberia.

When the isolation of Hindustan from the rest of Asia began, we do not know. Here, neither early Carboniferous nor Devonian plants are known. At the end of the Permian period, the connection between the Indian and the Angara flora begins to become noticeably lively. In the Angara flora, glossopteris and, perhaps, some other Gondwanian plants appear. In the Triassic period, these connections are already quite obvious. In the Jurassic period, the ways for the migration of plants were even wider. In India, a tropical flora of European appearance appears, although it has a local flavor. The isolation was over. Approximately the same picture is visible on the rest of the Gondwana continents (Carcaillet, 2007).

SUMMARIZE

For all this reasoning, the reader could lose the logical thread. Therefore, let us summarize.

1. In the late Carboniferous-Early Permian times, the position of the floristic boundaries of the highest rank, and therefore of the climatic zones, was sharply asymmetric with respect to the equator. Throughout the Southern Hemisphere to the South Pole inclusive, and in the north even to Kashmir there was an extratropical climate. The entire tropical zone fits in the Northern Hemisphere and forms an uneven strip with a strong strain in the region of India. The northern extratropical zone is 10-15 times smaller than the southern one.
2. At the present time, the asymmetry of the climatic zones of the Earth disappears if we single out (as was done for the Carboniferous-Permian time) only two types of climatic zones: tropical and extratropical. The modern tropical flora lies in a symmetrical band with respect to the equator.
3. It is impossible to explain such asymmetry of Paleozoic climatic zones by currents and other external factors. On the example of Eurasia, we saw that the floristic (and therefore climatic) boundaries of higher rank are controlled by a more general cause - solar radiation.

If so, then the following questions arise, addressed to opponents of the movement of continents.

1. Why did the Sun, during the entire existence of the glossopterya flora, warm less than one of the hemispheres? The displacement of the poles can not help here, since the hemisphere will remain so in any position of the earth's axis. This can be seen by picking up the globe, not the map. The boundaries of the Paleozoic flora allow deviations of the pole by 10-15 ° towards the lower reaches of the Lena, but not more. Otherwise, the pole enters the region of the tropical flora and the belt of the latter is located meridionally. It is pointless.
2. Why did the northern border of the tropics stay for almost 250 million years in almost one place, while the southern border was moving steadily southward?
3. Why was there a paleobotanical wall that fenced off India from the rest of Asia? Why a few hundred kilometers (and even with the archipelago of islands) did not allow the exchange between the Indian and Angara flora? After all, if you adhere to the theory of bridges and other fixist views, even thousand-kilometer spaces are not an obstacle to establishing a community of flora.

Well, it seems, now locked the fixers to the wall, now they will not turn away! In fact, we will look at the map. How tempting to take and move India to the south, to connect the Gondwana continents! Then imagine the entire mobilistic history of the Earth and admire how everything fits well with paleobotany ...

How good it would be if a simple and understandable picture was always correct, and a witty explanation of the paradox by a complete theory! But fixism also has its own trump cards.

Most of the above arguments are well known to the fixers, although two facts have become known very recently. This, first, is much larger than expected isolation of the Gondwana flora and, secondly, a high degree of independence of the most important floristic (climatic) boundaries from currents, mountain ranges and other external factors.

Opponents of the movement of the continents are well aware of the asymmetry of the climatic zones of the late Paleozoic. True, while they always talk about the shift to the north of the equator and do not take into account (and sometimes may be silent) that not only the equator, but the entire tropical zone is located in the Northern Hemisphere. It does not focus on the unevenness of the tropical zone. The unusual location of cold-loving plants and glaciations in India was tried to explain by the fact that these plants grew on the mountains, here also formed glaciers. However, this contradicts the presence in the thickness of the glossopterian flora of the present marine deposits.

In a protracted dispute, the Fixatives do not so much respond to the reproaches of the mobilists as they put forward observations that are not explained in terms of mobilism. Of these, perhaps, only one relates to paleobotany (we will not stay on geological and geophysical objections). In the middle of the Atlantic Ocean, 1000 km west of Africa, about 10 ° N. In the contents of the tube, which takes soil samples from the ocean floor, numerous shells of freshwater diatom algae and the remains of grasses were found. You can understand the joy of supporters of the theory of bridges, when they became aware of this fact. Of course, one can not exclude the possibility of drifting all these remnants with currents moving from the continent, but equally it is possible to assume the former existence in recent geological times of a rather large land mass in the Atlantic Ocean far from the coast of Africa.

It is possible to put forward one more consideration in favor of a possible constancy in the location of the continents. We do not know how the climatic zones would be distributed on the Earth if the Indian and Atlantic oceans are drained, and instead of them, Gondwana is placed as a single giant continent. Reproduction of such a climate model is a matter of the future, and without it is difficult to say the last word in a long-standing dispute. It is impossible not to ignore the fact that when solving climatic problems for the past geological epochs, we look at the modern appearance of the planet, willingly or unwillingly. However, we already had a chance to make sure that this image is not permanent for the Earth. This should never be forgotten.

In recent years, our popular scientific literature clearly shows a tendency not to create the reader's view that complex and controversial issues have already been resolved. It is much more correct and, of course, more fair to show the vicissitudes of the dispute and all the problematic nature of the proposed solutions. It is better to do this with the hypothesis of moving the continents. Of course, paleobotanical facts are in good agreement with it. However, it would be reckless only on this basis to hand over the whole fixist concept to the archive. Researchers who stand on the nobilist positions are still far from peaceful life, and it is not known whether a victory awaits them ahead. Will it not be so, that mobilism and fixism will still be able to form together the platform on which the third fundamental theory will be created, convincingly answering all tricky questions?

REFERENCES

1. Fisher, G. (2011). Kingdoms or dynasties: Arabs, history and identity before Islam. *Journal of Late Antiquity*, *4*(2), 245–267. https://doi.org/10.1353/jla.2011.0024
2. Rothwell, G. W., & Nixon, K. C. (2006). How Does the Inclusion of Fossil Data Change Our Conclusions about the Phylogenetic History of Euphyllophytes? *International Journal of Plant Sciences*, *167*(3), 737–749. https://doi.org/10.1086/503298
3. Archangelsky, S. (1996). Aspects of Gondwana paleobotany: gymnosperms of the Paleozoic—Mesozoic transition. *Review of Palaeobotany and Palynology*, *90*(3–4), 287–302. https://doi.org/10.1016/0034-6667(95)00088-7
4. Herrera, F., Manchester, S. R., Jaramillo, C., MacFadden, B., & da Silva-Caminha, S. A. (2010). Phytogeographic History and Phylogeny of the Humiriaceae. *International Journal of Plant Sciences*, *171*(4), 392–408. https://doi.org/10.1086/651229
5. Carcaillet, C. (2007). PALEOBOTANY | Charred Particle Analysis. In *Encyclopedia of Quaternary Science* (pp. 1582–1593). https://doi.org/http://dx.doi.org/10.1016/B0-44-452747-8/00187-3
6. Mendoza, R. U., Beja, E. L., Venida, V. S., & Yap, D. B. (2012). Inequality in democracy: Insights from an empirical analysis of political dynasties in the 15th Philippine Congress. *Philippine Political Science Journal*, *33*(2), 132–145. https://doi.org/10.1080/01154451.2012.734094

8 A HUNDRED-YEAR DISPUTE OVER A PINE CONE

From all that has been said, especially from the previous two chapters, the reader may get the impression that the paleobotanist is asking too many new questions, almost without answering the old ones. This, of course, is true, but not to the end. The contribution of paleobotany to the knowledge of the geological structure of vast territories - and, above all, of coal-bearing basins - is exceptionally large. All new emerging issues mostly relate to the most common patterns in the history of the planet and its flora. Here, indeed, even more unsolved than the known. However, many important studies have been carried out in this direction, which led to a clear understanding of many general laws (Ceballos, & Alvarez, 2013). The history of one such study we will consider. It will be a pine cone, which at first glance may be of interest only to foresters, squirrels and lovers of samovars. It is not the first time an object that a biologist chooses for his study causes a skeptical smile on the part of orthodox fighters for an early practical application of scientific research. How many sarcastic epithets were appropriated to the fruit fly of the Drosophila, this beloved pet of geneticists. Pine cone, the nature of which botanists have argued for nearly a hundred years - is also a worthy object for irony. However, we will not laugh at her, remembering that before she cast a complex piece of cast iron, her model is made of more pliable material. In addition, who knows, do not choose botanists for analysis just the pine cone, maybe the whole evolution of conifers would be incomprehensible to us.

ESSENCE OF THE QUESTION

Let us make the bumpkin the same as the protein does with it, and cut off the scales from it. There will be a bare axis, on which scars of scales located in a spiral order are visible. This spiral is clearly visible on the whole bump, especially if it has not yet opened. Now we take a separate scales (Fig. 33, d). It is dense, woody, on top with grooves for seeds, so it is called "seed." The seeds themselves are equipped with a verruca. Beneath the scales on young cones, a much more thin flake attached to the axis is visible. This is a covering scales. It covers the seed from below (DiMichele, et al., 2008).

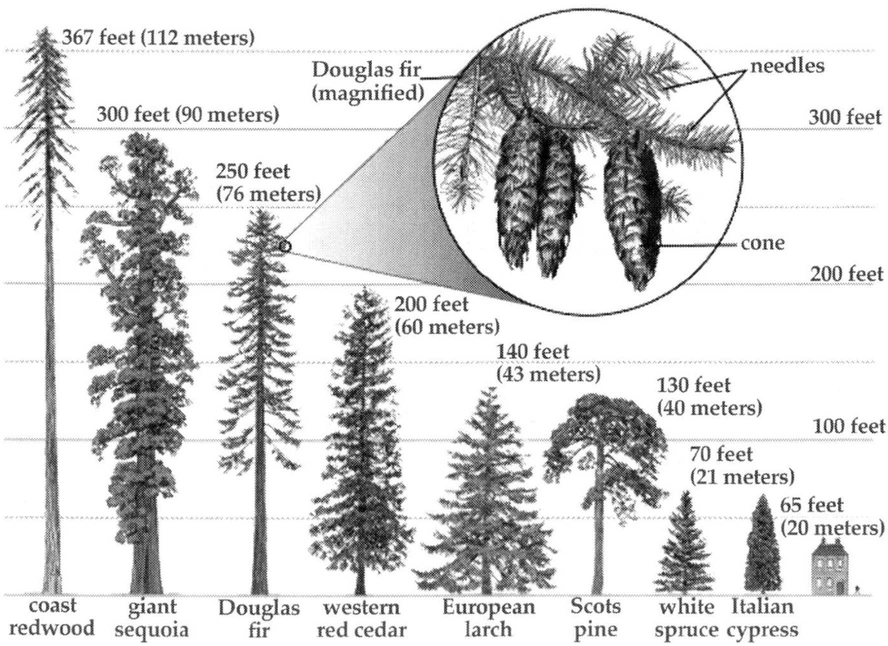

Figure 33: Evolution of the female escape from the Cordayites to the coniferous.

Now look at the lump of lepidodendron. Here, too, there is an axis, and also scales sit spirally. On each of them sits on top of one large sporangium. In some sporangia small microspores, in others - large megasores. Both those, and others, when ripe, were sown from cones, fell to the ground and, if everything was good, sprouted. We obtained small plants (sprouts) with sexual organs. Men's sprouts grew from microspores, and women's from megaspores. Sometimes megaspores sprouted even in sporangia. Thus, the megasores are the future female spasms (Landry, 2006).

Compare the cones of lepidodendron and pine. Outward similarity is definitely there. Both here and there spiral scales, on them the organs of reproduction. Pine has seeds, lepidodendrons have sporangia with megaspores (or microspores). It can be fully imagined that there was some similar to the lepidodendron cone, then it became more complicated and from the sporangia with the megasores the seeds were obtained (Ceballos, & Alvarez, 2013). We received in a greatly simplified form one of the concepts of the origin of pine cones, or rather, cones of modern conifers in general.

However, here for further reasoning we need to introduce the concept of "homology." This is better done by example. Suppose a certain mammal gradually connected its life with the sea more and, finally, finally passed to the water way of life. The usual paws became useless, and after a while they were replaced by fins, the seal's seal, the human arm and the leg of the elephant are not very similar to each other, but nevertheless these are organs of the same origin, they are homologous. Nature does not like to spend energy on the formation of a new organ, if an organ that has become obsolete is at hand, and it can be altered. Studying the laying and development of organs, it is possible to establish homologies. The results of such studies are sometimes unexpected. In fact, try to imagine that bird feathers are homologous to scales of holes on the snake's body. It is difficult to imagine that the stomata of a plant are a homologue ... an escape. However, Indian paleobotanist D. D. Pant recently showed that, most likely, this is exactly so.

The establishment of homology - alpha and omega morphology of plants, and indeed any morphology. When the pupil learned that in the words "toy" and "win" the common root, he also understood his homology. Knowing homology, we see the general after the particulars, we clarify the origin of both individual organs and organisms as a whole. Sometimes homologies are established easily (flippers and arms), sometimes it takes decades. To understand the pine cone, it took almost a hundred years.

The German botanist A. Braun proposed a correct general solution of the problem in the last century. "It fantastically combined the idealistic natural philosophy and inductive method of investigating the naturalist, rampant romance and spiritualism and a strictly scientific approach to the phenomena studied." This is how A.Takhtadzhyan, a prominent Soviet botanist, characterized A. Brown. Brown's ideas of interest were developed by the Czech morphologist L. Celakovsky (Ceballos, & Alvarez, 2013). According to the theory of Brown-Celakovskii, the seed scale of pine cone (the dispute is mainly about its nature) is the homologue of the whole shoot, but it is greatly shortened. This comparison, perhaps, is no less unexpected than the comparison of luxurious plumage and scaly skin. It is necessary to have a fair amount of courage, so that a branch with leaves and all adaptations for reproduction should be compared with one gnarled scales. The interpretation offered by Brown and Chelakovsky can only now is considered a theory. For a long time it was based on the analysis of modern plants, it was not confirmed by paragraphs of the paleontological chronicle, and therefore was only a witty hypothesis. Omitting further excursions into the field of modern plants and other theories about the seed scales, let us return to paleobotany.

IN THE BEGINNING THERE WAS A CORDAI

Many times we have discussed the Cordayites and now we return to them. Let us recall their appearance: large trees with thick trunk, spreading crown, leaves or short (as in lily of the valley), or long (as in iris), penetrated by parallel veins (Figure 18). Between the leaves sit earrings, which now need to be acquainted closer. This is a thin axis, and on it in two rows - small bumps sitting in the sinuses of long processes (Figure 33, a). Each knob consists of a short central axis and a large number (several dozen) of scales arranged along a regular spiral. The closer to the top of the knob, the more often you can see the legs with seeds in place instead of scales. Each leg has one seed. In other bunches on the same legs, several sacks with pollen (Ceballos, & Alvarez, 2013). The earrings of the Corday-tov were long, sometimes up to 30 cm. Thus, the French paleobotanist B. Reno understood the structure of the Cordaite earring when he obtained an extensive collection of mineralized remnants of Cordayites from the Oten coal field. Here there were not only earrings, but also trunks, leaves, roots. Renault thoroughly figured out everything and wrote an article published in 1879. The illustrations to it still go from the pages of one textbook or reference book to another. Note, by the way, that the Reno laboratory was located in a large wooden box in front of the doors of the luxurious entrance of the Paleontological Museum in the Botanical Garden in Paris. In such a construction, a talented and hard-working scientist worked all his life. It was a rare enthusiast.

It is still difficult to compare the cordaite earring with a pine cone, and there are many differences between the cordaite and the pine tree: long leaves with numerous veins at the cordait and bunches of needles in the pine. However, in a completely different way, a comparison is obtained if we take not the pine, but its ancestors.

When there were conifers, we do not know. Like all other large groups, they tried on their outfits somewhere behind the scenes for a long time and appeared before the paleobotanists already dressed in uniform, although somewhat old-fashioned. The first reliable conifers we meet in the last third of the Carboniferous sediments in Europe (including ours in the Donbass) and North America (Ceballos, & Alvarez, 2013).

The peculiarities of burial of conifers in rocks suggest that they appeared on higher elevations and therefore were not initially preserved in a fossil state. Confirmation of this was obtained by the Kiev paleobotanist AK Shchegolev on the Donetsk material. Looking at the general appearance of conifers, you can agree with this. Their entire structure is quite suitable for the settlement of well-lit and poorly moistened places. The leaves of the conifers are small, covered with a dense cuticle, and underneath it there is often also a mechanical tissue, giving the leaves strength and weakening the bright sunlight.

European conifers of the Carboniferous and Permian periods belong to several genera, but for us now only lebachia is important (Lebach is a place in the Saar). If the cones of the Lebachians were found separately from the branches and even in gutted form, it would be quite easy to compare them with the earrings of the Cordayites. They also have small knobs, which consist of a bundle of scales attached to the axis in a spiral fashion. Only the number of seeds is much smaller and sometimes reduces to one. Sits are in the sinus of a flat flyer, similar to a fork for fish and covering the scales of modern conifers (Figure 33, b). As if the first analogies with them are beginning to be outlined. All the ancient coniferous cones are spiral on the axis. This brings them closer to modern conifers, but distinguishes them from the Cordaites (we will return to this difference). In addition, should not we now try to compare the Lehbachian flyer with the covering scales, and the axis of the earring with the axis of the pine cone?

The study of conifers, younger than lebachy, showed Florina that this is exactly how homology should be established. He studied a successive series of coniferous Permian and Triassic periods and saw less and less scales remaining in the bump, how gradually they merge with each other into a flat plate and finally take the form of a flat seed scales (Figure 33c, d), On which depressions for seeds are formed. In addition, the covering scales remained almost unchanged (Nixon, 1996). Interpretation of individual parts of pine cones is no longer in doubt. However, this does not mean that all issues have already been lifted.

WHEN DID THE SEEDS APPEAR?

In addition, in special articles, and in more general works, the seeds of Paleozoic plants are often spoken of. When we just talked about the lump of the lebachy and the cordaite earring, we also said "seeds." They are found in cones, in ear rings, and just scattered in the rock. They did not doubt for a long time that these were really seeds, they had such a characteristic appearance: a lionfish, like an elm, a dense nucleus in the middle, a leg. However, even here the skeptics found what to doubt, and, apparently, they were right.

To understand the cause of skepticism, we will have to turn briefly to modern plants, at least to the same pine. For example, it is better to take the Siberian pine, which is more popular under the name "Siberian cedar." Speaking of ferns and lepidodendrons, we talked a little about their life cycle. These are spore plants. In sporangia, which are located near the ferns directly on leaves, and lepidodendrons in cones, there are disputes. Ferns are all the same, and lepidodendrons have two types: micro- and megasores. Spores are sown, and out of them grow sprouts. Of the same spores of ferns grow bisexual sprouts. All the plants with different spores and sprouts are different - male and female. Spermatozoa fertilize the ova, from which the habitual plant again develops. Such a life cycle is common to higher spore plants. Where is the pine all this?

It has sporangia with micro- and megaspores, and microspores (pollen) in each sporangium are many, and megaspores - one by one. Already in some spore plants, including lepidodendrons, some archaic ferns (we recall the biscuit, from Chapter II) spores germinate right in the sporangia. Pine and other gymnosperms are no longer an exception, but a rule. Pollen of a pine is a small, not fully developed male. Megaspory also germinate directly in sporangia and develops before the formation of oocytes. Megaspory in this state, with all the shells, are called ovules. It is not a seed yet. The seed will only be when the pollen gets to the ovipositor, the male adipose will end its development, the egg will be fertilized and the embryo will grow from it. When all these processes come to an end, you can gnaw a nutlet and extract a core from it, which is actually intended for feeding the germinating germ.

Now we find out that all Paleozoic seeds, which have managed to study the internal structure, have a megaspora that has germinated, and do not have an embryo. In other words, it is not seeds, but seeds. Pollen penetrated them even when they were on a tree, and we can often see it inside the ovaries. The embryo, apparently, developed much later, and when - we do not know (by the way, the embryo also does not immediately develop in modern ginkgo) (DiMichele, et al., 2008).

Paleobotany is well aware of all this and still call seeds - especially if they are not found in cones or earrings, but just scattered in the breed. Therefore it is excusable to do this to us. However, we must remember that it is not yet known when these “seeds” became real seeds.

We deliberately ignored the issue of the origin of ovaries, but this requires new tedious excursions into the field of plant morphology. It is better not to let them in. We will touch upon this issue in the next chapter.

BURIADIYA IS CONFUSING MAPS

In 1966, I happened to visit the laboratory of prof. D. D. Panta in Allahabad University (India). In difficult conditions a group of paleontologists worked: close rooms, photographic paper and chemicals are bought with their own money. But from the pen of these intuzia-stov every now and then there were articles written at a very high scientific level. Just in the autumn of 1966, D. Pant and his closest disciple, D. D. Nauthil, completed the study of the "coniferous" Buriadia (quotes, as we shall see below, are not accidental). The article was soon published under the modest title "On the structure of Buriadia heterophylla (Feistmantel) Sewardet Salmi and its fruiting", but there is no doubt that paleobotanists dealing with a wide variety of plants will apply to it (Schwark, 2002).

In 1879 O. Feistmantel found samples of conifers, very similar to the volcia from the Triassic sediments of Europe, among the samples collected by British geologists in India in the Giridikh coal deposit (Figure 34). To this kind of conifers Feistmantel and carried the Indian leaves. Later they were identified as a special kind of Buriadia, but no one doubted that it was coniferous.

Figure 34: Plants of volcia and Buriadia.

It is possible to imagine the astonishment of the Allahabad paleobotanists when they found seeds from the Buriadia that sat on thin legs right between the leaves without any cones, earrings and similar clusters. This can be expected in seed ferns, in some new primitive group of plants, but not in a plant, all resembling a typical coniferous species. Approximately the same feeling can be experienced by a fisherman if he finds a fish, in which the caviar does not lie in the belly, but hangs all over the body (Manchester, 2001).

Pant and Nauthyla, naturally, had a suspicion that this was not at all coniferous. A large amount of material was at hand (his Alahabad paleobotanists gathered in an expedition organized with funds from his own pocket), and everything that modern paleobotanical equipment can provide was taken from the samples. Cuticle preparations were prepared from leaves and seeds, separate seeds were separated from the rock with the help of hydrofluoric acid, the organic film was pulled from the rocks with special mixtures, pieces of stems were passed through acids, removing all solid, encased in celloid and cut like modern plants. Pantu and Nautilus were able to consider almost everything that paleobotany needed - and the carrina was completely entangled. According to the anatomical structure of the branches and the epidermal structure of the leaves of the Buriadia, the closest to the coniferous. The leaves are sometimes forked, and occasionally even more complex, in the form of a short spatula with denticles at the end. This also can be found in ancient conifers. But the pollen, which had already penetrated the seeds, was completely incomprehensible. Such pollen has ginkgo and bennettital. Inside the seeds of Buriadia, the layers of the cuticle are arranged, like no other gymnospermous plant. We already spoke about how the seeds were sitting. In one word, it was worth digging the buriadium deeper, and instead of all clear coniferous plants, there was nothing like a hybrid on the light (Manchester, 2001).

Having weighed the pros and cons, Pant and Nauthil decided that the best way to regard the burria as conifers is of course coniferous (of course, very peculiar, bearing the features of the ancestor of the Lebakhia). They abandoned the point of view of R. Florin that conifers originate from the Cordaites. In their reasoning there is logic, but only future research will show who is right. In any case it is obvious that in Buriadia there are features inherent in different groups of plants, and it clearly can not be a single ancestor for all. Consequently, these features could develop in her independently of other groups. The interpretation of the structure of the Gondwana Buriadia therefore raises an interesting question about the independent appearance of various types of conifers. However, this idea is not new. As early as 1940, she was touched upon in one of her articles by R. Florin. Based on the analysis of modern and fossil conifers, he came to the conclusion that the Gondwana conifers during the geological history were held by an independent group, which is felt to this day. In this respect, the discovery of Pant and Nautiyl agrees well with Florin's ideas.

WHAT'S LEFT FOR THE SKEPTICS?

The evolutionary series painted by R. Florin, leading from the oldest conifers to modern ones, is perhaps the most beautiful in all palaeobotany. Paleobotany built no less accurate rows for other plants, but there were always some inconsistencies. That lacked one or two links - and instead of them hypothetical forms were drawn, the descendant was older! Ancestor. Beginning with Lebakhia and ending with modern conifers, a phylogenetic branch is emerging, in which serious flaws are not yet visible. However, all that was with coniferous before, is quite problematic (Landry, 2006). In other words, the connection of coniferous and cordaite, planned by R. Florin, still causes many doubts. True, to doubt that they are close relatives, yet do not have, they have a lot of similarities. The wood of the ancient coniferous and cordaite is one and the same, their pollen and seeds are similar, and even those "bumps" (as we found out, they are shortened shoots with seeds) are also surprisingly similar. But the doubts remain like this.

First, it is necessary to withdraw needles of conifers from long large leaves with numerous veins. However, this question can still be answered. The ancestors of conifers clearly had leaves with more complex veins. Not without reason, in some of the most ancient conifers, we find individual leaves with a forked apex. Congenital and conserved ability of conifers to give complex leaves is manifested in some modern and fossil genera of different families. For example, the leaves of agathis (dammaras) are externally almost like the Cordaites. All this - so far only logical reasoning, and plants with foliage, intermediate between coniferous and cordayites, we do not know. But the second question is much more complicated. All paleobotanical data show that the spiral arrangement of leaves, shoots, propagation organs in evolution precedes a two-row arrangement. Why should conifers be an exception, and how their cones with a spiral arrangement of scales could come from the earrings of Cordayites, where the bumps sit in two rows? There can be only one answer to this question: most likely, the conifers did not originate from those Cordayites that we know, but from older, more primitive forms (Landry, 2006). It may also be that both these large groups have their origin from some common ancestor. The solution of the riddle of the origin of the first conifers will come from very ancient (maybe Devonian) sediments. It is possible that such an ancestor was one peculiar coal-bearing plant found in Scotland, in which the wood is cordaite-coniferous type, and the leaves are scaly-like, approximately, as in araucaria.

REFERENCES

1. Landry, J. T. (2006). Dynasties: Fortunes and Misfortunes of the World's Great Family Businesses. *Harvard Business Review*, *84*(11), 36. https://doi.org/10.1111/j.1540-6563.2007.00197_74.x
2. Manchester, S. R., & Tiffney, B. (2001). Integration of Paleobotanical and Neobotanical Data in the Assessment of Phytogeographic History of Holarctic Angiosperm Clades. *International Journal of Plant Sciences*, *162*(S6), S19–S27. https://doi.org/10.1086/323657
3. Schwark, L., Zink, K., & Lechterbeck, J. (2002). Reconstruction of postglacial to early Holocene vegetation history in terrestrial Central Europe via cuticular lipid biomarkers and pollen records from lake sediments. *Geology*, *30*(5), 463–466. https://doi.org/10.1130/0091-7613(2002)030<0463:ROPTEH>2.0.CO;2
4. DiMichele, W. a., & Gastaldo, R. a. (2008). Plant Paleoecology in Deep Time. *Annals of the Missouri Botanical Garden*, *95*(1), 144–198. https://doi.org/10.3417/2007016
5. Nixon, K. C. (1996). Paleobotany in cladistics and cladistics in paleobotany: enlightenment and uncertainty. *Review of Palaeobotany and Palynology*, *90*, 361–373. https://doi.org/http://dx.doi.org/10.1016/0034-6667(95)00092-5
6. Ceballos, F., & Alvarez, G. (2013). Royal dynasties as human inbreeding laboratories: the Habsburgs. *Heredity*, *111*(April), 114–121. https://doi.org/10.1038/hdy.2013.25

9 PARALLEL CURSOR

We've gotten acquainted with enough facts to reflect a little now. When you read books or articles on the general problems of the development of the organic world, it is always evident how little factual material extracted by paleobotany is used for generalizations. Of course, this can be explained by the fact that "fossilized plants" are not a very grateful object for deriving the laws of evolution. But, most likely, the matter is that the paleobotanists themselves, who are overburdened with work on the preparation of preparations, determining the collections collected by geologists, systematizing the continuous flow of new and incomprehensible plants, simply do not find the time to present their observations in a form suitable for generalization. It should not be forgotten that a qualified paleobotanik must also be a good geologist at the same time. It is difficult for him to grasp the immensity. However, this same encyclopedic activity enabled some paleobotanists to become outstanding theoreticians. They are G. Potonie, E. Jeffrey, DG Scott, O. Linier, A. N. Krishtofovich, A. L. Takhtadzhyan and others. Their works shed light on many common problems of biology.

In this chapter we will examine the phenomena of parallelism. It's about independent similarities. When we see identical wheels from different cars or a vacuum cleaner in the form of a rocket, we are not very surprised. Man so needed, it was convenient (nice, profitable, etc.), he thought - and did so. When we find the same details for different plants, about "thought and did", of course, there can be no question. Explaining such phenomena, they often say: "In certain circumstances, the relevant structures have emerged." This, of course, is not an explanation, but a kind of demagogy. In the second half of this century, we need to show concrete causes and mechanisms. Therefore, we will start with concrete examples.

SKILLFUL SIMULATORS

The amazing similarity of far from not related plants we saw on the example of the Buriadia and the usual coniferous Northern Hemisphere.

Another example, which will be discussed now, is no less indicative. We have already met Gondwana and Angara horsetails, whose leaves are narrow and long, but have grown together in the base in the form of a bowl. This phyllotheks (in Greek "phyllon" - a sheet, and "teka" - a bowl). As early as 1828 the French paleobotanist Adolf Bronyar first described them on samples from Australia, but did not depict a single specimen. Then other researchers published images, and the genus widely entered the literature. In 1879 the Russian paleobotanist IF Shmalhausen described and depicted a whole series of species of phyllothes from the Kuznetsk, Tunguska and Pechora basins. His drawings were so vivid that it was the Angara phyllotheks that began to be cited in handbooks and textbooks when they wanted to show which plants should be attributed to this genus.

In the early sixties, MF Neiburg engaged in the Pechora phyllotecamp, drew attention to specimens with sporangia. With neat rosettes they sat between knots right on the stem. Sockets are pressed together and close a significant part of the internode. I wanted to understand the structure of sporangia, and MF Neiburg instructed me to do this work. Outlets of sporangia in phyllothes are small, several millimeters in diameter. The work was laborious. It was necessary to make consecutive sections every

25-40 microns. It took several days. After each cut, a photograph was taken and a sketch was sketched. In all, there were about two hundred slices. The results obtained at first failed to be used, since soon after the end of this work MF Neiburg did not become. In 1966, when I was comparing the Gondwana and Angara phyllothes, I returned to a pack of photographs and drawings made in my time and began to study them carefully. I was familiar with the work of the Australian paleobotanist D. Taunrow, who studied the sporangia of real Australian phyllothes. Externally, the shoots were almost the same, but the sporangia turned out to be quite different. In the Angara phyllothes, there are several rows of rosettes between the nodes, and in the Australian little above each node there is only one whorl of the branching legs (Figure 35). At the end of each branch, the scutellum (like the cap on the nail), on the back of which hangs four sporangia. One such shield can be compared with one rosette of sporangia in Angara species. However, on this similarity ends, Sections showed that the sporangia of the Angara phylloths had fused laterally, and the rosettes themselves sit on non-branching legs.

Figure 35: Spore-bearing shoots of Angara.

The reader, far from the systematics of plants, all these differences may seem not too significant. Therefore, let us hasten to explain that they are abundant enough to include the Angara and Australian (and with them and in general Gondwan) phyllotheses not only to different genera, but also to different families. True, here we immediately find ourselves in a mess: the sporangic sprouts of both are devoid of such similarity that we simply can not identify serious, for example generic differences between them. Here there is a difficult question with a generic name, but this is not what interests us. It turns out this way: two different plants belonging to different families (this is at least) with fundamentally different organs of reproduction have similar shoots (not only leaves or stems but whole shoots) that it never occurred to anyone that it is even different births. In the literature only the issue of species differences was discussed.

Of course, this similarity did not arise out of nowhere. Analysis of the evolution of arthropods shows that both the Gondwana and Angara species can have common ancestors. They have to be found somewhere in the lower parts of the Carboniferous sediments or even in the Devonian. Since that time, the evolution of phyllothes has been going on in the Northern and Southern hemispheres quite independently, but in parallel.

Other cases of parallelism, concerning not the whole plant, but only its leaves, we met earlier. Completely identical leaves with middle vein and almost perpendicular lateral veins (meaning the formal genus shadowioptery) can independently have some cicadas, bennettitales, marattia ferns, pteridosperms and a peculiar group of Gondwana plants - pentoxylenes. This can also get a ficus leaf, if it is found with dangling edges. An example of parallelism is given by the leaves of the Cordayites. Their taxonomy was rebuilt on the basis of the microstructure of the leaves. Then it turned out that earlier researchers who studied only the general form of a leaf, attributed to one species leaves belonging to different genera. An even more striking example is the external similarity between real and seminal ferns. We have already said that both of them originate somewhere in the Devonian from plants with primitive foliage, and then completely independently and in parallel come to a very similar external appearance.

WHAT DOES THE MICROSCOPE SAY?

The microscope always helps the taxonomist, helps him to understand the essence of phenomena. What will he say about all these imitations? The microscope has already answered one part of the question. It was thanks to the microstructure and the different nature of outwardly similar leaves was revealed. The other part of the question is much more complicated: if in some cases the similarity is independently formed in leaves, seeds, stems, that is, at a macroscopic level, then, in other cases, similarity can manifest itself at the level of cells and tissues? This question can be safely answered in the affirmative. Such cases are known, but how often it happens - we do not know.

In tissues that carry nutrients and water through the body of the plant, there are cells with different walls. If we trace the development of the stem of a highly organized plant, we will see how first the ringed and spiral thickenings form on the walls of strongly elongated conducting cells (tracheids), then the spiral begins to contract, its adjacent turns are connected by bridges, a ladder is formed, gradually, in the wall, only more or less difficult Arranged pores (Figure 36). Further evolution leads to the disappearance of the walls at their ends, so that continuous long tubes-vessels are formed. Such sequential complication of the wall of conducting cells is observed in many groups of plants, but the process stops at one or another stage. For example, in pine and all other conifers, it does not reach the vessels that are the privilege of flowering plants, in lepidodendrons development ends in the staircase tracheids, and in the case of spiral tracheids.

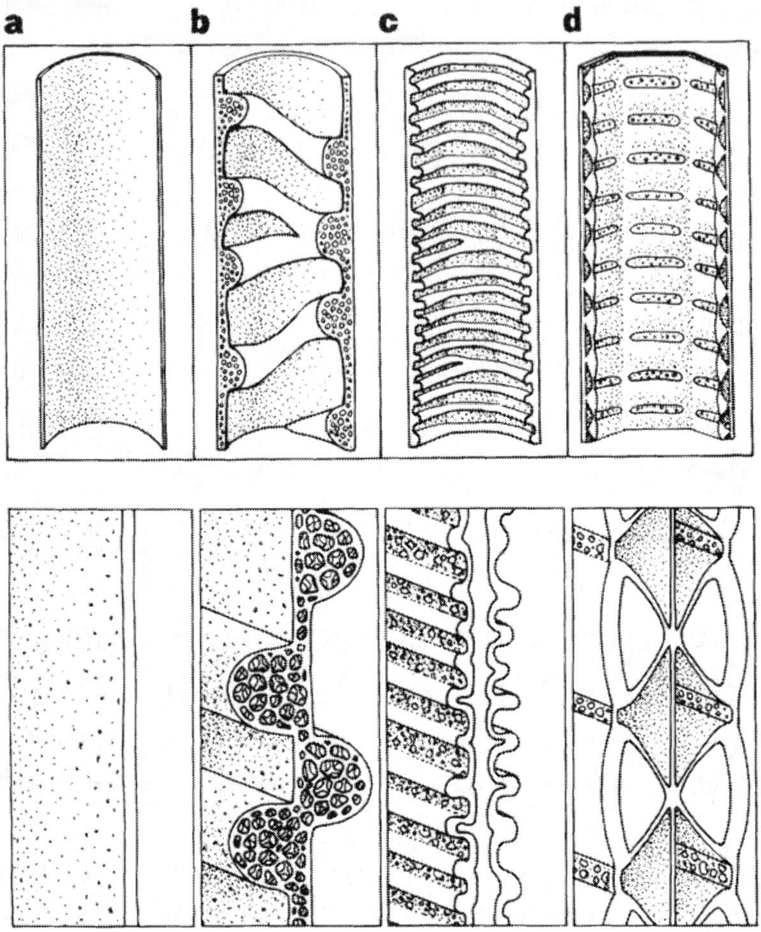

Figure 36: Development of the wall of the conducting cells (tracheid).

Now let's talk. Let's take two evolutionary lines. One, including the placenta, begins from very ancient plants known in the lower horizons of the Devonian system. They have only annular and spiral, that is, the most primitive thickenings. We leaf through the pages of the geological record and see how the complication of the wall of the conducting cells (tracheid) is going on in the evolutionary process of the placenta. There are staircase tracheids, and then the real pores. Up to this stage, evolution has already reached the Carboniferous period. For some modern, we find, even rarely, even vessels. The second same line starts from plants like rinia, and here we see the same sequence. If we knew more fully the evolution of arthropods, then, undoubtedly, we would see here the same course of evolution of the conducting tissue. In any case, some of the modern horsetail vessels are marked. Hence the conclusion: in the microstructure of plants a parallel, independent acquisition of the same characteristics is. This can be seen particularly well in the case of vessels that appear in highly organized placentas, horsetails, ferns, gymnosperms, and many angiosperms. Thus, in the "ability to imitate one another," microscopic elements do not lag behind leaves and stems.

Before proceeding further, we need to discuss some terminology. We have already said in Chapter II that the phenomena of similarity not reinforced by kinship are widespread in the organic world. At the same time, a series of terms were given to denote different types of such similarities. We will not explain all the terms, but on the two still stop: on convergence and parallelism. Convergence is usually said to be an outward resemblance of totally unrelated organs or organisms, which has emerged under the influence of similar external conditions or a similar way of life. This is the similarity of a dolphin and a shark. They also say that under convergence, the similarity is never deep, that is, it does not penetrate into the fine structure. It is possible, but not necessary, that concurrency is ultimately determined by similar external conditions, but we are talking about it when closely related groups that have a common ancestor develop independently in the same direction. So the zoologists reason. What are we dealing with in plants?

Here, these similarity phenomena are so widespread that we can not confidently divide them into related and not related deep and superficial, connected with external conditions and unrelated. Therefore, it is best for us not to make any distinction between convergence and parallelism and to unite all similarity phenomena under one term. So we will proceed, and according to the traditions that have developed in the botany, we leave the term "parallelism."

DECOMPOSITION OF POLYNOMIALS

Now I must make a warning to readers. It should have been done, perhaps, at the beginning of the chapter, but it's better late than never. The fact is that we are entering the field of theoretical botany and paleobotany. Therefore, I suggest that lovers of easy entertaining read the book no further and take up other books or magazines. And we, the most patient and inquisitive, will continue our journey among fossil plants.

So, we got to the phenomena of parallelism, noted the "insidiousness" of plants, which not only rip apart, but they also make the same fabric and make up their tissues, but they did not say anything about the reasons. Here, of course, it's easy to get rid of a magical spell with a reference to the external environment, but we will try to understand the mechanisms and regularities. To do this, we use tried and tested methods.

When a researcher encounters a complex phenomenon, he first of all tries to decompose it into its component parts. A mathematician decomposes a polynomial, a chemist separates individual elementary reactions, a geneticist examines the variability and inheritance of individual characteristics. Similarly, obviously, we should do, analyzing the evolution of ancient plants.

The first who understood this and dismembered the plant into elementary organs (or rather, the elementary evolving units) was German paleobotanist G. Potonier. He described the mass of fossil plants, did extremely much in the knowledge of the Permian and Carboniferous deposits of Europe, laid the foundations of the modern doctrine of the accumulation of fossil fuels, wrote a mass of popular scientific articles, published a journal and, finally, made a huge contribution to the overall morphology of plants. G. Potonier died in 1915, when he was barely over fifty.

Potonier understood that the sheet is not a simple simple formation, but a complex synthetic organ, the result of the coalescence of elementary structural units. This idea was then developed by O. Linier at the beginning of this century. Linier distinguished several elementary units from plants. Some give rise to large leaves, such as ferns, and sporangia develop on them; Others represent the outgrowths of the first and are always devoid of sporangia (this, for example, leaves of the placenta); The third give rise to the roots. Proceeding from his constructions, Linier actually predicted the discovery of psilophytes.

Unfortunately, at the beginning of this century, few botanists thought about such problems. Therefore, the ideas of Potonier and Linier passed almost unnoticed for contemporaries. According to one of the modern botanists, they were "a voice crying in the desert." Only in the 30s the

German morphologist V. Zimmermann brought these ideas to their logical conclusion, widely propagandized them and provided their constructions with successful terminology. Simple and monotonous organs of the first terrestrial plants - so to speak, elementary evolving units - he called telomes. From the telomes that underwent various changes, a whole variety of external forms was formed in terrestrial plants. Zimmermann also decomposed these changes into elementary processes: fusion, bending, thickening, sterilization, etc., which provide the plant with better absorption of light and better protection of sporangia. Indeed, if the sporangium protrudes at the end of the stem, it is most susceptible to both temperature fluctuations and various mechanical damages. The bent leg already gives the sporangium an advantage, and even better, if it sits in the axillary of the leaf under its protection.

The fact that such elementary processes could occur in a variety of plants is beyond doubt, and paleobotanical observations confirm this. Take, for example, the moss-like and the arthropods. In their evolution during the Paleozoic and Mesozoic, the same processes are observed, leading to the formation of very similar structures. Sporangia, first scattered over the shoot in groups, then gather into compact cones at the ends of the stems. In these cones, the leaves are transformed into special formations that protect sporangia. The sporangia from the stem themselves pass to these protective formations. Similar processes take place in both groups and in the microscopic structure of the stems. It is possible that the mice and arthropods are connected by a common origin (there are ancient plants that reveal features of both groups), the more similar is the similarity in their evolution. This is one of the main sources of parallelism in development, but not the only one.

Every living organism, be it an animal or a plant, is not a simple set of organs, independently performing their functions. Changing one organ causes a chain reaction in the organization. The adaptation of plants to propagation of seeds by wind reduces the size of seeds, leads to changes in the structure of the ovary; The development of a new group of muscles leads to a proportional strengthening of the corresponding region of the nervous system, etc. Such dependences are incalculable, and the more they are, the more highly developed the organism is. It is quite natural that the reaction of different organisms to the same change may turn out to be the same if the internal dependencies are also the same for them.

This is the second source of parallelism. Unfortunately, it is difficult to give illustrative examples, because for this it is necessary to climb into the terminological jungle.

From what has been said, we need not conclude that the organs do not have any autonomy and are absolutely necessary. In the process of evolution, they feel themselves completely independently. In plants, even in the most highly organized, the dependence between organs is much less than in animals. A typical, albeit somewhat crude example: if you tear the tail off the dog, you will not be able to grow a new dog from it, but from the branches of one tree, for example a poplar, you can grow a whole grove of normal trees. Because of the high independence of organs among plants, amazing creatures, combining archaic and progressive signs, which is found (a disharmonious plant, for example, is magnolia). From this it follows that the relationship between the organs is unlikely to have played a leading role in the parallel evolution of different plant groups.

The third source of parallelism, apparently, lies in the genetic possibilities of living beings. Genetics is now actively popularized, with its terms more and more are getting acquainted with the general public. But just in case, we give two definitions. We need the terms "genotype" and "phenotype" in the arguments. The genotype is all hereditary information that the body receives at the time of its inception. Throughout life it is not used completely, but only to the extent that environmental conditions permit it. The result of the interaction of hereditary information and environment, that is, the organism itself, is called a phenotype.

In the course of evolution, plants often simplify their structure. This is proved, for example, for the spore-bearing bodies of the Carboniferous Klinolists. In the course of such simplification, usually occurring through a delay in the individual development of the organ or its parts, the former complexity remains in the genotype. In it, as it were, the possibility of the subsequent development of complex structures, if necessary, is preserved. And if such an opportunity suddenly occurs in the branches that have already diverged from the simplified ancestor, then it is not surprising that similar signs are formed. However, we are touching upon a problem that is still very little developed. The available illustrations concern only modern plants, and paleobotany can not help much here.

Genetic nature has also the parallelisms described by NV Vavilov in his classic classical series. Although he traced parallel variability in modern crop plants, the patterns he derived are more general. For a long time these studies were outside the field of view of biologists and, unfortunately, did not find any response in paleobotanical generalizations. Therefore, it makes sense to stop on homologous series with regard to fossil plants.

The essence of the regularities discovered by Vavilov in brief is reduced to the following. It is quite obvious that the more closely related the organisms are, the more common they are. This is the starting point of biological taxonomy. At the heart of each species is a certain set of mandatory attributes, without which we must speak about another form. But other signs, which are not species, show a greater or lesser degree of variability. Studying numerous varieties of cultivated plants, NI Vavilov noticed that this variability is ordered. In closely related species of the same genus, it goes in almost identical directions, and the closer the species are, the more completely this similarity. The point is that the difference in the form of seeds (or other organs) in species belonging to different genera becomes smaller than in species of the same genus (Figure 37). It turns out a mass simulation of organs. The same phenomenon is observed in the analysis of families, orders, etc. But in fact the limited variability means also the limitations in the assortment of characteristics. Plants can not go beyond a certain range of features, although the circle is very wide.

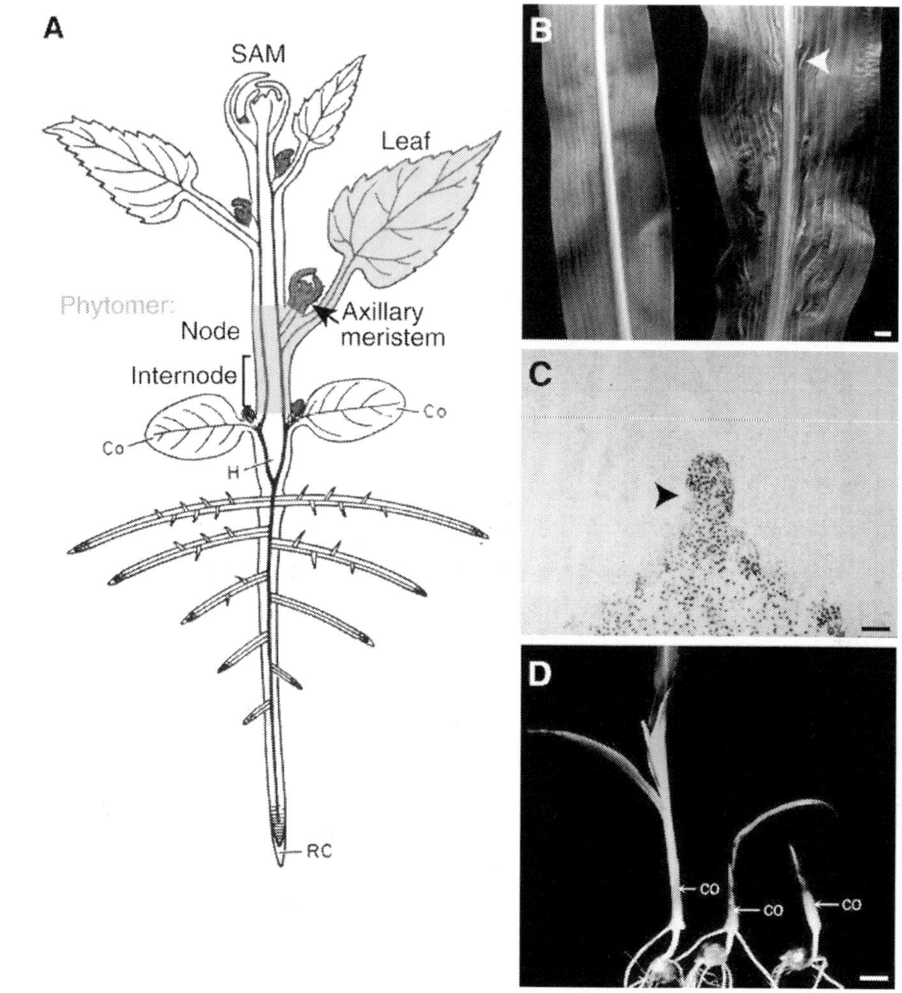

Figure 37: Parallelism in the structure of leaves in wheat.

Examples of such limited and recurring variability among systematic groups of different levels are clearly visible among fossil plants. When it was a question of cordaite leaves, there were two of their anatomical types, behind which are different genera. In one genus (Rufloria), the stomata are hidden in the grooves, while the other (Cordaites) are scattered over the leaf surface. On the basis of the microstructure, the second genus can be subdivided into several subgenera. Within each genus and subgenus, it is possible to find leaves of the most diverse appearance: long and short, narrow and wide, with a rounded and sharpened apex, etc. The microstructure is a permanent stable feature, and external features are subject to considerable but uniform variability. Therefore, classifying leaves according to outlines, density of venation and similar signs without taking into account the microstructure, paleobotanists included in one species leaves actually belonging to different genera.

Before we conclude these long arguments, we note that the listed ways for the appearance of parallelism are not only probable, but also proven. However, we can not say with which of them in each specific case we are dealing. The reason for this is our almost complete ignorance of the adaptive value of the most basic signs of a plant. Perhaps the most interesting and incomprehensible is the independent appearance of very fine microscopic structures in tissues, the appearance of surprisingly similar spores or pollen in a variety of groups somewhere in the middle or at the end of their evolutionary development. What we are dealing with here is completely incomprehensible. Of course, we are very accustomed to treat the microstructure somewhat with greater respect than to the general morphology. Researchers easily admit the formation of the same form of leaves in unrelated plants, and having seen similarities in the structure of tissues - are inclined to immediately see kinship. However, we have no special reason to believe that in the nature workshops, jewelers are at a higher price than carpenters, and that they are not forced to do the same work several times.

REFERENCES

1. Calvillo-Canadell, L., & Cevallos-Ferriz, S. R. . (2002). Bauhcis moranii gen. et sp. nov. (Cercideae, Caesalpinieae), an Oligocene plant from Tepexi de Rodríguez, Puebla, Mex., with leaf architecture similar to Bauhinia and Cercis. *Review of Palaeobotany and Palynology*, *122*(3–4), 171–184. https://doi.org/10.1016/S0034-6667(02)00135-5
2. Taylor, R. S., Tarver, J. E., Hiscock, S. J., & Donoghue, P. C. J. (2014). Evolutionary history of plant microRNAs. *Trends in Plant Science*. https://doi.org/10.1016/j.tplants.2013.11.008
3. Gibling, M. R., & Davies, N. S. (2012). Palaeozoic landscapes shaped by plant evolution. *Nature Geoscience*, *5*(2), 99–105. https://doi.org/10.1038/ngeo1376

4. Banks, H. P. (1987). Comparative morphology and the rise of paleobotany. *Review of Palaeobotany and Palynology*, *50*(1–2), 13–29. https://doi.org/10.1016/0034-6667(87)90038-8
5. Rudwick, M. J. S. (1972). *The Meaning of Fossils: Episodes in the History of Palaeontology. History of science library.* https://doi.org/10.2307/2407183
6. Piperno, D. R., Moreno, J. E., Iriarte, J., Holst, I., Lachniet, M., Jones, J. G., … Castanzo, R. (2007). Late Pleistocene and Holocene environmental history of the Iguala Valley, Central Balsas Watershed of Mexico. *Proceedings of the National Academy of Sciences of the United States of America*, *104*(29), 11874–11881. https://doi.org/10.1073/pnas.0703442104

10 UP THE SECTION

In this chapter, we will skip the Devonian, Carboniferous and Permian flora, which are sufficiently described in the previous chapters. The reader should already have a general impression of the plant world of the Paleozoic. In the Mesozoic, when the famous dinosaurs came on stage, the plants were completely different: ginkgo, coniferous, cicadas, bennettitales, and then angiosperms, came to the fore instead of treelike plaques, numerous seed ferns and cordaites. On the flora of the Triassic, Jurassic and Cretaceous periods, we will talk a little lower, and now we will try to understand how the scenery changed, where all the plants that settled the land in the Mesozoic "hid"?

THE ROOTS OF THE MESOZOIC FLORA

It is no accident that the word "hid" was used. As in earlier periods, at the boundary of the Paleozoic and Mesozoic we do not see traces of the workshop where new genera and species are being developed. But if we can not grasp their origin, then at least look at their appearance.

Many great artists first worked as extras, participated in mass scenes, and then they seldom had the good fortune to even make a retort. With the same inconspicuous precursors of the Mesozoic flora, we begin to meet quite early (Somerville, & Koornneef, 2002). The first reliable ginkgoids are known even in the Lower Permian sediments of Europe. The first plants with foliage, as in cicadasses and bennetti species, appear even earlier, in the upper part of Carboniferous deposits. At the same time, conifers begin to gain strength. At the end of the Carboniferous period, these plants were almost invisible against the general background of the tropical flora. In the first half of the Permian period their number increases, and at the beginning of the second half of this period, when the basic plants of the Carboniferous forest have already died out, their places become very numerous.

In the last century, the paleobotanists carefully studied the plant remains (Figure 38) in the copper shale of the GDR and West Germany, lying at the very base of the Upper Permian stratum. These shales are still being developed. They found only a dozen species: mostly conifers, ginkgo and ancient cicadas, there is one kind of pteridosperm, almost no ferns and horsetails. From the former splendor of the Carboniferous forest, there is no trace left. And the appearance of plants, and the rocks in which they are buried, everything speaks of a hot and dry climate. It is possible, of course, to assume that this particular flora, which clearly has Mesozoic features, gave rise to new rich forests of the Triassic, Jurassic and Cretaceous periods. However, by the end of Perm, the set of species in it got smaller and smaller, it seems that the flora of the copper schists has died out, leaving not too abundant offspring. In addition, the plants of cuprous schists, accustomed to a rather severe hot climate, could hardly change quickly and turn into moist forests of the Mesozoic. In addition, they probably had worthy competitors, already accustomed to wet conditions.

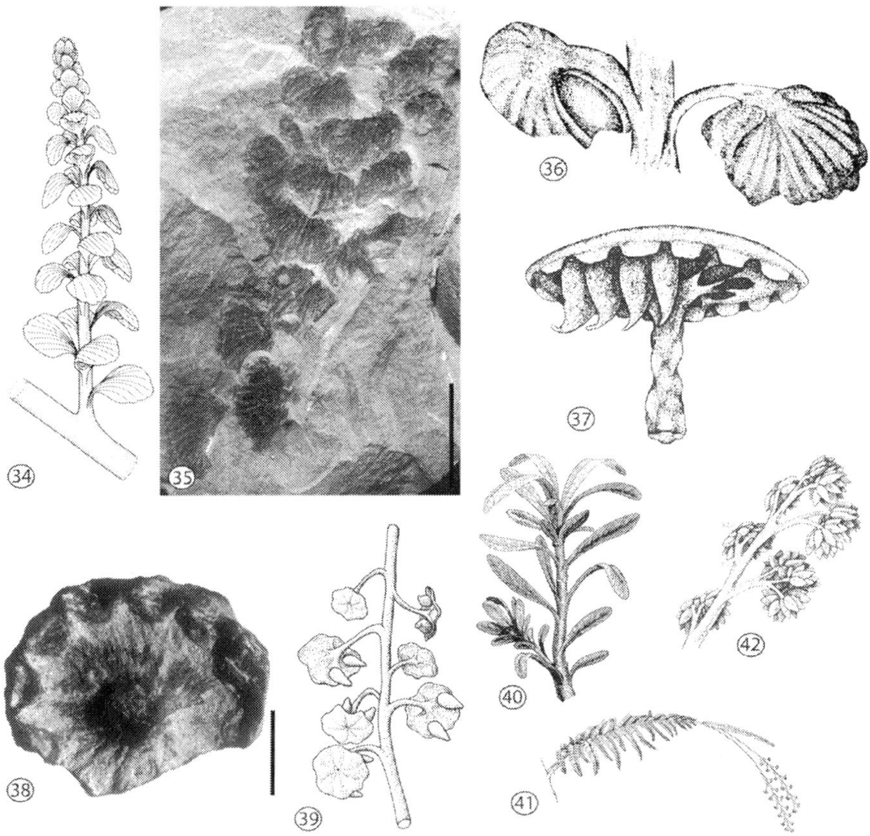

Figure 38: Plants of Upper Permian copper schists of Western Europe: Seed Fern Lepidopteris.

There really were competitors. At the very end of the Carboniferous period in Europe, the drying of the climate began, which progressed throughout the Permian period and captured even a part of the early Mesozoic. However, the farther east we move, the less and less noticeable are the traces of dryness. In the northern part of Eurasia Angarida was located with its extratropical flora. Between it and Gondwana in the territory of China, Korea and South-East Asia was located one more ancient continent Catania with a tropical flora. Here the humid tropical climate lasted for a long time. Therefore, some of the plants of the Carboniferous Forest survived almost to the end of Perm. We are not able to follow the belt of moist tropical vegetation of the Permian period far to the west. But to assume its existence somewhere south of the drylands can be. Here are some observations.

At the beginning of the 1960s, an interesting collection of fossil plants was found in the Upper Permian sediments of the eastern part of Turkey: European ferns of the rock-coal-Early Permian type, Gondwana glossopterites and some Kata-Asian plants. With such an amazing mixture of shapes, paleobotany faced for the first time. But, maybe, there simply were mistakes in definitions? No, hardly. This flora was determined by the English paleobotanist R. Wagner, one of the best connoisseurs of European Paleozoic ferns. Certainly, he certainly defined them correctly! Gondwana glossopterists can not be confused with anything. Catatrical plants are also not less than catchy, characteristic species. One of them is gigantopteris. It is a fern-like plant (maybe a seed fern). Its merged feathers form a single leaf, sometimes very similar to a leaf of tobacco or alder.

Two more facts are interesting. First, the Catalan plants are found in some places of Gondwana, namely: in South Africa and South America. Secondly, in the west of the USA - in Texas, Oklahoma and Colorado - the Permian flora was met, which was first directly called the Catalan, and then softened the formulation and began to speak only about plants of the Catalan type. The discovery of the Catalan appearance of plants in North America, whose flora as a whole developed according to the European pattern, made a great impression at the time. And the mobists, and fixers, that is, both supporters and opponents of the movement of the continents, were puzzled. After all, both agree that the Pacific Ocean already existed in the Paleozoic (or perhaps it was an age-old one). Continents moved or not, but in any case plants to get from China to the southwest of North America, had to cross the thousand-kilometer oceanic space. The thought of the researchers began to look for possible bridges. They can not be thrown from the north and from the south: here is the domain of domination or Angara or Gondwana extratropical flora, through which the tropical inhabitants can not pass. It is assumed that the similarity of Western American and Chinese plants shows parallelism, of which we have already spoken. An interesting explanation, but it requires confirmation by a detailed study of the plants themselves. This work is a matter of the future. Apparently, everything can be explained differently.

We saw a map of the modern vegetation of the humid tropics. It extends along the equator with a solid ring. Why not assume the existence of the same belt in Perm? Why the humid tropical forests in the Permian period should have been inhabited only by the Catalan continent? They could well be to the west and pass between the flora of the arid hot climate in the north and the Gondwanian flora in the south, that is, through Asia Minor, the Middle East, North Africa and Central America. Along this belt with a moist tropical climate, the Catalan plants freely walked westward, mixed with the Gondwana (in Asia Minor), and themselves penetrated the Gondwanian flora (in South Africa and South America). Finally, some plants, which had changed considerably after long wanderings, could rise from this belt to the southwestern part of North America. Here they founded a colony, the remains of which are confusing paleobotanists. This, of course, is only a hypothesis based on the discovery of the Catasian appearance of plants away from the places familiar to us. For its confirmation, more detailed and numerous facts are needed (Somerville, & Koornneef, 2002).

However, it's time to explain why we paid so much attention to this belt. After all, we discussed before this the question of the origin of the Mesozoic flora. The fact is that in the Catalan flora there were many different plants of the Mesozoic appearance from the very beginning of the Permian period, and maybe even earlier. By the end of this period, the number of such plants increased. This rejuvenated Catalan flora, accustomed to tropical conditions and settled on the hypothetical belt, which we talked about a lot, could well become one of the sources of the heat-loving Mesozoic flora of both hemispheres, give it cicadas, ginkgoids, and some ferns.

PALEOBOTANICAL POMPEY

In the role of inhabitants of Pompeii, plants were often found. To much they were able to adapt; they adapted to salt water, to deserts, and to high mountains. But here "take into account" that they did not know that suddenly a lava will flow from the neighboring mountain, and clouds of ash will fall from the sky! As time went on, the volcanic cloak again settled first the most unpretentious, then more finicky plants. Again the soil was formed, and life continued as usual. Then everything was repeated again. Volcanic rocks with plant remains are of great value for paleobotany. It's not just that the plants are sometimes preserved in them no worse than in coal kidneys (Figure 39). The main thing in the other. Above it was said that the plant can be preserved for centuries only when its parts fall into the water. For this, the plant needs to grow near water bodies. Volcanic rocks can bury a much wider range of plants. Maybe that's why imprints in volcanic deposits are especially incomprehensible, peculiar, and therefore interesting.

Figure 39: Mineralized stalk of fern in volcanogenic deposits of Siberia. The surviving cell structure is well visible.

At the very end of the Permian and early Triassic, the whole territory east of the Yenisei and west of Lena, north up to Taimyr, and to the south almost to Krasnoyarsk was a real volcanic country. From it we left many meters frozen streams of lavas, huge strata of volcanic tuffs and rocks, formed as a result of the erosion of both. It is very difficult to understand the complex alternation of such breeds, especially in Siberia, where sometimes there are not a single outcrop on hundreds of square kilometers. But this must be done, since the most important minerals are concentrated in the volcanic strata of Siberia, among them optical raw materials and non-ferrous metals, oil was recently found here. In such cases, the geologist looks with hope at the remains of animals and plants, as indispensable landmarks in the geological context (Somerville, & Koornneef, 2002).

The study of the volcanic deposits of Siberia began only in the 1930s. Then the outstanding Soviet paleobotanist VD Prinada described the collected plant remains. His book was published only in 1970. Many of his other manuscripts were unlucky. They remained unpublished. VD Prinada died in 1950, while still in his prime (he was born in 1897).

The study of the Siberian paleobotanical Pompeii began very closely only in the mid-fifties. However, it proved to be a tough nut to crack, even for geologists who had seen it. Thousands of specimens with fossil plants have already been collected, the remains of various ancient animals have been found, but the question of the age of the sequence remains open to this day. His decision rests on very contradictory "testimonies" of plants and animals. Here is an example.

Already in the first chapter it was said that the microscopic shells of spores and pollen are much more frequent in mountain rocks than ordinary plant remains. The test sample is crushed, treated with reagents, the pellet is centrifuged several times and finally, these membranes are isolated. They can be seen on a fine microstructure, according to which they are classified. Such palynological studies in the Soviet Union are handled by several hundred people working in geological administrations and research institutes. When the first difficulties arose in the knowledge of the volcanic stratum of Siberia, the geologists decided to resort to this tried and tested method. Among the first were the Moscow geologist and paleobotanist G. N. Sadovnikov, who extracted the samples, and the palinologist (that is, the expert on fossil disputes and pollen), O. P. Yaroshenko. However, instead of clarity, only new questions arose.

Between paleobotanists of various profiles, there are often misunderstandings about the age of this or that stratum. One proposes to draw a line between the systems here, and the other - a little higher or lower. But such contradictory data as it turned out in Siberia, the history of paleobotany probably still did not know. Under the volcanic stratum here are coal-bearing Permian deposits. In them, there are good preservations of plant prints, spores and yyltsa. All these residues are well studied. In volcanic deposits, the picture changes dramatically. Even a person who is completely uninitiated in paleobotany can say that there is an exceptionally sharp change in the composition of the fossil flora. As through the earth numerous plants of the typical Cordaite taiga fall through, and on the stage without the slightest warning come various ferns accompanied by cicadas, ginkgoids, conifers and other plants. In other words, a flora appears that has a typical Mesozoic appearance. True, some Paleozoic forms among freshwater mollusks and crustaceans pretend that nothing has happened, and continue to exist quietly. They can be forgiven: after all, fauna is a fairly sovereign kingdom of nature. The most surprising is not this, but the fact that the controversy and pollen in the breed also "reacted badly" to the events. They, as it were, resolutely separated themselves from those who, according to the logic of things, produced them. Of course, there are no miracles in the world, but we cannot say why there are no fundamental changes in the composition of the prints of plants so decisively on the one hand, on the one hand, and the dispute, pollen and animals on the other (Somerville, & Koornneef, 2002). We do not know until now the level at which the boundary of the Paleozoic and Mesozoic passes in Siberia.

In any case, we have one of the oldest and perhaps the richest Mesozoic flora. The works of recent years, in particular the observations of GN Sadovnikov, show that there are many plants here that were previously considered characteristic only for the Jurassic and even the Cretaceous periods. It is possible that volcanic eruptions buried exactly those plants that somewhere in the depths of Siberia were preparing for a wide outlet into the future Mesozoic forests. The question of the origin of the plants by this assumption, of course, is not removed, since we again see not some ancestral, but quite developed forms. But even interesting is that paleobotanists approached the oldest representatives of the Mesozoic flora.

In the paleobotanical literature, the question of the specific origin of the Mesozoic flora is sometimes discussed. They are looking for a certain Ararat, to which the botanical Noah's Ark has moored. Meanwhile, each significant in time and distribution in the area of flora could well appear from the mixing and interaction of people from many places. In particular, the formation of the typical Mesozoic flora could simultaneously make its contribution to Siberia, and Catasia, and the Urals, and plants found in European copper-schist, and Gondwana. When climatic barriers began to break down in the Triassic, a massive "fraternization" of flora from different regions began.

ON THE MEANING OF SMALL THINGS

A systematic geological study of the Pechora coal basin began as early as the late 1920s. By the end of the 1950s, a lot was known about it, detailed geological maps were compiled, hundreds of wells were drilled, coal and oil deposits were intensively developed. It is understandable that geologists were mainly interested in coal-bearing deposits, although they paid enough attention to what lies below and above them. And yet in 1958 it turned out that in the course of the work the deposits of the whole period were omitted. On the geological map, the Triassic deposits were not mistakenly indicated. Actually, they were seen and described, but did not know that they were Triassic. On the basis of general geological considerations, the young geologist AL Yanshin (now an academician) spoke about the possibility of such an age in 1936. But his assumption was then left almost without attention.

In 1958, in one of the wells near the Bolshaya Synya River in sediments considered Permian, the geologist FI Yentsova discovered the remains of plants that were transferred to the definition of MF Neiburg. Just at this time, after graduating from university, I came to work at the Geological Institute and one of the first days I saw Mikhail F. Neiburg talking with the geologist A. V. Macedonov, under whose supervision Ms. Entsova worked. I saw the impression made by Neiburg on Macedonov that the Triassic flora found on the Big Son, that the plants from this place are well known in the Upper Triassic sediments of Central Europe. For geologists working in the Pechora basin, this was a big event. I had to reconsider my views, change the geological map, look again at these deposits, trace them, collect the remains of plants and animals (Massini, et al., 2010).

Among the samples from the Big Son, there were several leaves with a well-preserved coal crust and fairly distinct veins. Imagine a fern, whose end feathers are fringed, and veins are still located, and you will get the impression of the plant found here. True, it was not an ordinary, but a seed fern. Reading the literature, MF Neiburg found images of very similar leaves from the Southern Urals. They were attributed to the genus Aipteris, which was considered a close relative of the Cata- ciatic genus gigantopteris (it was already discussed in this chapter). Gigantopteris is a Permian plant, and Aipteris was first found in Permian sediments. Numerous finds of Aipteris in the Triassic of the Pechora Basin and in the Southern Urals suggested that the flora of the Catasia (Massini, et al., 2010) directly affected the formation of the Early Mesozoic flora of these places.

For gigantoptery and its closest Kata-Asiatic relatives, the formation of reticular venation is characteristic. This means that the small veins join together and cover the entire plate of the leaf with a fine mesh (Figure 40, a). The same connection of the veins appeared to be seen on the leaves of the Aipteris from the Pechora Basin and the Southern Urals. But then there were doubts. Geologists have found in the Pechora Basin, and then in Bashkiria, leaves that have undergone a natural chemical treatment. In a flat bag of cuticle, all veins are perfectly preserved here. And then it turned out that they never connect with each other (Figure 40, b). Now it is good to say "it turned out." Behind this word is not easy work on attentive study of prints and phytolum, analysis of literature, special microscopic technique. Together with the preparation of cuticle preparations, this work was taken from the Moscow palaeobotanist IA Dobruskina for more than a year.

Figure 40: Venation of the seed fern from the Triassic sediments.

When such a conclusion was made, it was necessary to abandon the kinship with the Kataazi gigantopteris. It was necessary to look for relatives again. They played a role of a fraction of a millimeter, to which veins did not reach each other (Massini, et al., 2010). Do not think that for a plant these fractions of a millimeter are a trifle. Fused veins immediately lead to a different course of metabolism in the plant, and this can not be neglected.

Fortunately, their relatives were found pretty soon. These were the same leaves from the Central European Triassic on which I. Bornemann hundred and a half years ago for the first time in the world began epidermal paleobotanical studies. The best examples of these leaves are kept in the Swedish Museum of Natural History. In response to I. Dobruskina's timid request to send from the doublet fund, at least for a while, for comparison a slice of coal crust or a cuticle preparation from Stockholm came a weighty pood box with the very leaves described in the literature.

"The Sistine Madonna" is known to everyone, in millions of reproductions it has spread all over the world. Still, the best reproduction never replaces the original. Therefore, not out of courtesy, there was a crowd of people at this picture, when the Dresden Gallery was exhibited in Moscow. Perhaps it would be an exaggeration to say that a paleobotanik experiences the same feeling when he sees an imprint in front of him, on a photograph or picture of which he looked many times in the book. But a similar feeling, of course, arises. Finally, you can most consider all those details that have been omitted or simply not noticed by your predecessors, to check the previous conclusions, both your own and others'.

The samples sent from Stockholm finally persuaded Dobruskin that the closest relatives of the Pechora and South Urals leaves lived in Europe, and not in Catasia. In the biography of the Early Mesozoic flora, a well-documented record appeared. The game played some kind of a trivial little thing. But on such little things and keeps modern paleobotany. Conversely, neglecting them sooner or later will lead to a gross paleobotanical error, and the chain reaction that has begun will eventually strike the interests of geological practice (Massini, et al., 2010).

HARRIS SCHOOL

In 1963 the paleobotanists of Moscow and Leningrad accepted the English paleobotanist, a member of the Royal Society (in our opinion, an academician), and the president of the Linnaean Society, TM Harris. The work of this English researcher must be told specifically. Harris did not make great paleobotanical discoveries. Nevertheless, in the paleobotanical summaries and textbooks published in the last 30 years, information and reproductions of illustrations to his works invariably appear. It can be said without exaggeration that the appearance of Harris's works marked a new stage in paleobotany. And it's not that he introduced any ingenious methods to paleobotany, but that from the material chosen for research he extracted everything that is possible.

Sometimes in a collection that is the basis of a published article or monograph, later, interestingly uncovered interesting fingerprints are found in greater or lesser numbers. It happens that there is a beautiful sheet on the piece of rock, and next to it there is some kind of nondescript snippet. He did not even pay attention to the paleobotanik who studied the collection. The same material gets to a more thorough researcher, and then it turns out that really "not everything is gold that glitters." A fragment is more valuable than a handsome man. Few of the paleobotanists extract as much from the material as Harris and his disciples (Massini, et al., 2010).

This style of work is visible even in the earliest articles of Harris. In the 1920s he occupied the Triassic-Jurassic flora of Greenland, and in the first monograph, published in 1926, strikes the full extent of the coverage of the material. Some specks on the phytolumus attracted his attention - and he found out that these were the remains of mushrooms unknown to science. Unclear prints, barely visible on the breed, - after careful study, were found to be well preserved remnants of hepatic moss, also previously unknown. But this attention to detail was not enough for Harris. It is not enough to see what lies on the surface of the rock cleavage. How to look at what lies inside? Harris is developing a new method called "bulk-maceration." In a literal translation, this sounds rather awkward: "bulk maceration", or "maceration of volumes." A large piece of rock is placed in a mixture of nitric acid with bertholete salt and is destroyed (macerated). The sediment is passed through the sieves, and at the disposal of the paleobotanical are various [small leaves, scraps of cuticles, sporangia, seeds and a mass of other interesting material.

A careful study of the samples allowed Harris to establish new, hitherto unknown groups of plants. One of them is leptostrobus. This genus was described in the last century by the Swiss paleobotanist O. Heer, when he was studying the collections collected in the Jurassic deposits of the Irkutsk basin.

Together with leptostrobuses, there are often peculiar thin leaves collected in bundles. This - Chekanovskaya, a genus named after the Poles AL Chekanovsky, a famous explorer of Siberia. In appearance, the leaves of Czekanowski were long considered to belong to the ginkgo.

How many times has this paleobotanistic "outward appearance" come down! After the lesson, taught by seed ferns, paleobotany seemed to be more cautious. But the principle formulated by DG Scott: "very similar - it's not the same thing" is struggling to make its way. Reason is once again proved by the history of Chekanov.

As early as the beginning of the 1930s, the Japanese paleobotanist S. Oishi, studying the cuticle of Chekanovsky, expressed doubt that it was ginkgoose. Later, Harris, having studied the cuticle and Chekanovskii, and leptostrobus, came to the conclusion that both paleobotanical genus actually belong to the same plant. Suspicion of Oishi was confirmed. A careful study of the leptostrobus showed Harris that this is a female cone of a very unusual design. On the long axis pair of scales, having the appearance of the shells of the shell, sit in pairs. Inside these valves there are several seeds. Ginkgo and his relatives have nothing of the sort (Figure 41a, b). Here the seeds either sit on legs separately, or are located on a short, branchy shoot. The peculiarity of plants with cones leptostrobus and Chekanovskaya foliage is such that the Indian paleobotanist D. D. Pant (who understood the structure of the Buriadia) considered it possible to distinguish them into a new order of gymnosperms, which he called Gzekanowskiales. Order is a very high systematic category. Suffice it to say that one order is either all conifers, or all ginkgo, or all cicadas (Taylor, et al., 2009).

Figure 41: Ginkgo.

One such study would be enough to take a firm place in the history of paleobotany. But this is only a small part of what Harris did. He extracted interesting information about the ancient cicadas, figured out the organs of reproduction of a peculiar group of catetonia (sometimes they are considered the ancestors of angiosperms), described Many new genera, gave an exhaustive description of the Jurassic flora of England, which has now become the standard. Speaking of Harris' studies of the Upper Triassic-Lower Jurassic flora of Greenland, American paleobotanist Andrews wrote in his textbook paleobotany: "There are few paleobotanical studies, if any, that would shed light on so many, hitherto unknown plants."

Harris was never taken for generalization on a planetary scale. Perhaps it was due to the fact that he, like no one else, knew the value of the facts, which all seemed indisputable. Working on the flora, previously considered quite decently studied, he always uncovered a lot of mistakes and errors, based on the superficiality and negligence of researchers. But Harris was never just a collector of facts and a strict professional auditor. After all, the unraveled nature of a previously unknown or incomprehensible fossil plant is not a naked fact, but a ready-made generalization and the necessary source material for a qualitative reconstruction of the vegetative kingdom of the geological past. Therefore, the work of Harris is not a file of factual material, but irreplaceable and reliable reference books. Subsequent researchers almost never found errors and ill-considered places in them. For these works, the Royal Society elected TM Harris as its member (Taylor, et al., 2009).

INEXHAUSTIBLE OPPORTUNITIES

We have already seen that when studying fossil plants and when investigating crimes, sometimes the same methods are used. This does not exhaust the similarities in the work of paleobotany and criminalist. Both of them, according to insignificant evidence, have to reconstruct past events, appearance and manner of behavior of the actors, to dig up to the causes of the phenomena.

Modern paleobotanical technique allows you to extract a huge amount of information from stone material. Unfortunately, these truly inexhaustible possibilities are usually not fully used. And it's not just that the paleobotanik does not have the necessary conditions for this, but also that he does not always know how to get down to such work. Sometimes he formally and superficially handles one collection and already takes the next one. This attitude to the material can be compared with the ill-considered exploitation of a mineral deposit, when only "cream" is removed from it. After such exploitation, the deposit is of little interest, although the richest deposits remained untouched in it.

Not so long ago, this approach was the only one possible, since in the interests of geological practice it was necessary at least initially to inventory the main fossil flora. But now from such studies, sometimes more harm than good is obtained.

Fortunately, among paleobotanists, from time to time there are people who, at their own peril and risk, undertake the re-licensing of such material. This is what Harris did when he took up the study of the Jurassic flora of England. In the same direction worked and his students - D. Tauirou, SM Arkhangelsky, W. Chaloner and others.

In our country, detailed studies have also been carried out, extracting from the material all possible information. As an example, let's take early Cretaceous plants of the Far East. During the last ten years, VA Krasilov studied them in the Vladivostok palaeobotanik. Prior to him, the basic inventory work had already been done. On Krasilov's share, the ungrateful role of the auditor fell out. It was necessary to adopt modern paleobotanical equipment. Here are some results of this work: more than 40 new species are described, in 12 species the names were incorrect, the systematic position of many plants was revealed. If we consider that about 140 species are known in the Lower Cretaceous deposits of Primorye, Krasilov's revision is quite significant. But this flora was considered well studied!

On the leaves of many plants, VA Krasilov discovered fruiting bodies of various fungi, which our paleobotanists did not study at all. It can not be said that they are not seen. Just did not attach any importance to some dark spots on the cuticle. Meanwhile, the study of the remains of fungi significantly complements the characteristics of ancient plant communities, allows the introduction of biostratigraphy a new group of plants.

Krasilov managed to extract disputes from the sporangia of many ferns. It has already been said that paleobotanists extract spores and pollen scattered in the rock. Sometimes this is the only way to find out the age of the rocks. Which genera or families belong to these microscopic remains are usually difficult to say. True, sometimes there are some catchy signs, according to which the researchers determined the systematic belonging of the spores and pollen. For example, it seemed that it was quite possible to recognize among the fossil spores the representatives of the Gleichenian and Shiza ferns. Such disputes were described in the composition of the corresponding families, and even special monographs were devoted to them. Again, external similarities led paleobotanists. Krasilov's works showed that disputes that were considered belonging to the family of Gleichenia could well belong to four other families. Equally unproved were the definitions on the spores of genera of other families.

THE KINGDOM OF ANGIOSPERMS

The middle of the Cretaceous period was a turning point in the history of the vegetable world. If in the Lower Cretaceous sediments we find the remains of ordinary Mesozoic plants - cicadas, bennettitales, ginkgo-vich, conifers and ferns, then in the Upper Cretaceous rocks the picture is completely different. Almost everywhere, we begin to meet numerous remains of angiosperms. With a sharp change of scenery, we have already encountered in previous chapters, but the advent of the century angiosperms occurred especially unexpectedly and largely incomprehensible. The famous Soviet botanist Mikhail I. Golenkin believed that at that time cosmic phenomena occurred that changed the nature of the earth's atmosphere and made it more transparent. A stormy stream of light energy could really be resisted only by angiosperms (Looy, 2013). This popular hypothesis was abandoned by paleobotanists. Most paleobotanists now come to the conviction expressed by VA Vakhrameev that angiosperms originally appeared on elevated and mountainous spaces. In the event of the emergence of extensive arid zones that led to the extinction of Mesozoic moisture-loving plants, angiosperms spread widely around the globe. In this sense, it is interesting that the most primitive of the modern angiosperms (for example, magnoliaceae) have preserved their attachment to mountainous places to this day.

The hypothesis of the mountain origin of angiosperms captivates with its logical nature, but does not solve all the questions facing paleobotany. It is not enough to answer the question "where", it is necessary to show still "how" they occurred Paleobotanical "documents" did not, unfortunately, preserve the corresponding "records." The most ancient angiosperms known to us already have all the main distinctive features of this class. Like all other major groups of the vegetable kingdom, the angiosperms appear already quite decorated. Therefore, the question of "how" should be answered not by paleobotany, but by botany (Farrell, & Mitter, 1998).

The main weapon angiosperms in their struggle for existence, of course, was a flower. But the decisive role was played not by the colors of the corolla, but by the structure of the ovary. Ovary consists of individual carpels, and in them are securely hidden from drying out and eating insects germs of future seeds. The emergence of this reliable system is the object of long-standing disputes between botanists and paleobotanists. Almost all the main groups of plants, even horsetails, were in the possible ancestors of angiosperms.

An unexpected and ingenious solution to the problem was proposed by the Soviet botanist and paleobotanist AL Takhta-jyan. He reasoned like this. The ancestors of angiosperms were scattered throughout the leaf. Sitting in the non-opening buds, such seed leaves had to be folded. Imagine now that the leaf remains in a folded (conduplicate) state, its edges grow together, and the seeds hidden inside continue to develop. As a result, you get an organ that can not be called a carpenter otherwise. Analysis of the flower of modern angiosperms showed that this could be the case. This ability to sexual maturation in an underdeveloped or simply embryonic state is called neoteny. The hypothesis of neotenic origin of angiosperms is shared, if not all, by so many botanists (Farrell, & Mitter, 1998). In the evolution of angiosperms, neotenic phenomena appear to have played a very large role. With these phenomena AL Takhtadzhyan also connects the origin of herbaceous angiosperms.

The victorious procession on the planet angiosperms began, being trees. It is not by chance that the most archaic among them, for example the magnolias, are always tree-like. According to the common opinion of botanists, grass appeared on the earth later. Herbaceous (especially annual) plants have serious advantages in front of trees. They have a short life cycle, early maturation. Being pressed to the ground, they are easier to tolerate the cold. In the harsh winter it is easier for them to hide under the snow. In herbs, generations alternate rapidly. This is also an advantageous feature that allows them to quickly change, adapt, and then spread over the surface of the planet. Only herbs were able to colonize high mountains, deserts and polar countries. Trees can not penetrate there. What is grass, what is it different from trees?

Let's follow the development of the tree. Here the seed sprouts, and a small plant is formed, called a sprout. Unlike animals, plant development proceeds by layering newer and newer layers of cells that overlap, but do not replace seedling tissues. The sprout itself, in fact, almost does not differ from the flowers deprived of flowers. In other words, the sprout, slowing down the development of all its organs, except for flowers, may well be considered as grass. It is appropriate to recall here that plants often rush to reproduce without achieving a normal adult appearance. In the botanical garden on the island of Trinidad, once unexpectedly blossomed mahogany, having a height of only 25 cm. Sometimes start to bear fruit very young seedlings of maple, pear, apple, elder and other trees. Usually it does not come from a good life. IT Vasilchenko discovered in Namangan a walnut stump with a one-year escape, which fruited. The seeds were sown and a new crop was obtained in the sowing year.

The conclusion that the grass is delayed in the development of trees, already suggests itself. This thought, now quite natural, but once unexpected, also belongs to AL Takhtadzhyan.

We did not say anything here about the facts obtained by the paleobotanists on the history of the flora, composed primarily of angiosperms. Such a flora inhabited the Earth, beginning with the second half of the Cretaceous period (Farrell, & Mitter, 1998). With the corresponding prints of plants, we meet on all continents. We will not consider all these documents, since this is not part of the task of our book. In such a young sediment, we, in essence, meet with modern genera, and even species. Some unusual plants are interesting only for specialists and hardly deserve consideration in a popular book. The story of the flora of the Cenozoic is inconceivable without getting acquainted with the systematics of modern plants, and this requires too much deviation in the paleobotany area. Nevertheless, I would like to note that in the decoding of this chapter of plant history a great role was played by the works of our remarkable paleobotanists, L. N. Krishtofovich, I. V. Palibin and their students.

The last millions of years passed under the sign of the growth of climatic contrasts on the Earth. At the beginning of the last, Quaternary period, ie, one and a half to two million years ago, the glaciation era began on Earth, which was a great event for the plant world. The displacement of climatic zones resulted in the migration of plants, extensive areas of the tundra and forest-tundra appeared. About the dramatic events in the life of plants in these difficult times for them, we learn mainly from the remains of the spores buried in the rocks and pollen. But this is already a very special topic for conversation. Those interested can read about this in SA Safarova's popular pamphlet "With a microscope deep into the millennia."

REFERENCES

1. Farrell, B. D., & Mitter, C. (1998). The timing of insect/plant diversification: might Tetraopes (Coleoptera: Cerambycidae) and Asclepias (Asclepiadaceae) have co-evolved? *Biological Journal of the Linnean Society*, *63*(4), 553–577. https://doi.org/Doi 10.1006/Bijl.1997.0207

2. Taylor, T. N., Taylor, E. L., & Krings, M. (2009). Introduction to Paleobotany, How Fossil Plants are Formed. In *Biology and Evolution of Fossil Plants* (pp. 1–42). https://doi.org/10.1016/B978-0-12-373972-8.00001-2
3. Massini, J. L. G., Jacobs, B. F., & Tabor, N. J. (2010). Paleobotany and Sedimentology of Late Oligocene Terrestrial Strata From the Northwestern Ethiopian Plateau. *Palaeotologia Electronica*, *13*(1), 1–51. https://doi.org/Artn 6a
4. Somerville, C., & Koornneef, M. (2002). A fortunate choice: the history of Arabidopsis as a model plant. *Nature Reviews. Genetics*, *3*(11), 883–889. https://doi.org/10.1038/nrg927
5. Looy, C. V. (2013). Natural history of a plant trait: branch-system abscission in Paleozoic conifers and its environmental, autecological, and ecosystem implications in a fire-prone world. *Paleobiology*, *39*(2), 235–252. https://doi.org/10.1666/12030

11 LIVING FOSSILS

Living fossils or organisms that are accustomed to see petrified, not alive, are a favorite topic of the popularizers of paleontology. And for a paleontologist it is always interesting to look at the appearance and way of life of an animal and a plant, from which only "horns and legs" are usually literally left. True, I must say that such findings did not lead to any important changes in our ideas. For palaeontologists in these beings, another is important. They need to know how long a biological species or species can survive. However, let's not jump ahead with reasoning. It is better to first look at concrete examples (Kenrick, & Crane, 1997).

"TREE FOR FORTY CROWNS"

The classical living fossil among plants is usually considered ginkgo. This is a large tree with a spreading crown and ornamental foliage. Such graceful bilobed leaves, intersected by a fan of thin veins and sitting on a long petiole, are not found in any other modern plant (Figure 41). Ginkgo is interesting not only for paleobotanists, but also for botanists. Together with the cicadas, it stands alone among gymnosperms by the way of seed formation. When the ginkgo pollen enters the ovum, a microscopic tube called pollen grows from it. In her, two spermatozoa appear shortly before fertilization. Equipped with cilia, they slowly move and resemble infusoria. Approaching the egg, the sperm leaves on its surface its spiral band with cilia and nudge penetrates inside. Sperm cells and eggs merge. Fertilization has occurred. In most other modern gymnosperms, spermatozoa are devoid of cilia and, consequently, of mobility. By the way, mobile spermatozoa of cicadas are larger than in all other higher plants. They reach a diameter of 0.3 mm and are easily visible to the naked eye.

Ginkgo seeds are also interesting. Strictly speaking, while they are hanging on a tree, they can not be called seeds. It's just a seed. Pollen in them falls in the spring, and fertilization is postponed until late autumn, when the ovules are already falling. It is possible that a similar picture was observed in many Paleozoic plants, for example in seed ferns, in seeds of which it was never possible to see the embryo.

Now ginkgo can be found in botanical gardens and parks in southern cities of Europe, Asia and America. Meanwhile, three hundred years ago, European botanists did not know anything about him. But in Chinese poems ginkgo figured back in the beginning of the 11 century, drawings and descriptions of it can be found in the Chinese scientific literature of the XVI century. Judging by Chinese sources, the seeds of ginkgo were taken in the north of the country in the form of tribute.

Probably, there are few plants with so many different and sometimes unexpected names. Dr. Kempfer, through which ginkgo entered Europe, was a physician at the embassy in Japan, studied the local flora and noted ginkgo in 1690. In the manuscript for ginkgo, three names were given: "ginan", "sankyo" and "icho." Already at the beginning of the 18 century. He combined the first two names together. As the Latin letter "g" appeared in the middle of the resulting word, it is still unclear. Perhaps it was just a mistake, and some paleobotanists, in spite of the established tradition, prefer to write "Ginkyo." In 1796 the Englishman D. Smith was indignant with the inconsistency of the combination of "kgo", called it barbarous and awkward, and proposed to call the plant Salisburia, but it was too late. For a quarter of a century before, Carl Linnaeus had already introduced the word Ginkgo into the official Latin nomenclature. From here it got into Russian and German. The English call ginkgo "Maidenhair tree", which literally translates as an absurd set of words "tree of girlish hair." The thing is that Maidenhair is the English name of the grassy fern adianum, or "venene hair." The feathers of this fern are similar to the leaves of ginkgo, and in search of a name for the last, the English turned to the fern, adding the word "tree" to his name.

The French call ginkgo very unusual: "a tree for forty crowns." This name, according to Loudon, was born like this. In 1780, the Parisian amateur botanist named Petigny visited London to inspect the botanical gardens. In one of them he saw five young ginkgo plants, which was so rare in England that the owner of the garden considered himself the sole owner of this plant. He himself raised trees from seeds obtained from Japan, and hoped to sell them profitably. However, after a good breakfast with the appropriate amount of wine, the owner of the garden gave Petigny all five copies grown in one pot, for 25 guineas. The guest immediately paid off and, without losing time, carried out the purchase. The next morning the sleazy owner found Petigny and offered him the same amount for one tree. The Frenchman was adamant and took ginkgo to his homeland. Each tree cost him 120 francs, that is, 40 crowns. Almost all the ginkgo of France, and there are many of them in this country, originate from the trees bought by Petigny.

These plants are rather unpretentious. According to one botanist, they survive in the worst environmental conditions that a human being can create. They are not afraid of insects and fungi. Perhaps, it is because of their extraordinary endurance that ginkgo survived to the present day (Kenrick, & Crane, 1997).

In China and Japan ginkgo is considered a sacred tree. It is possible that it was man's care that helped it become a living fossil - "something contrary to the ordinary fate of all life" (G. Endryos). In a wild state, these trees are known only in Southwest China, and even then there are doubts whether they have penetrated here from culture. Meanwhile, even in the Tertiary period, gipkgoes were distributed over a vast territory from England to Sakhalin. But the most abundant remains are found in the Upper Jurassic and Lower Cretaceous sediments. Sometimes the phytolamps of their leaves completely cover the surface of the stone or form entire interlayers. The most ancient imprints of ginkgo are found in the Upper Triassic sediments. These remnants are almost 200 million years old.

The role of ginkgo and his closest relatives during the Mesozoic times, that is, in the age of dinosaurs, was very significant, but still often exaggerated. Ginkgo is sometimes referred to as relatives of plants alien to him. Some of them, as now it became clear, are completely independent and, in any case, belong to other families and families. Others, for example, Czekanowski, generally found themselves in a completely different group of plants (see Chapter X).

Understanding the systematics of fossil leaves of ginkgoids is a very difficult matter. They are too changeable. Even on one tree, one can find leaves from rhombic to deeply dissected. Some leaves are even twisted into a deep funnel and fused laterally. In the rocks fall leaves from trees belonging to different species, and none of the species does not differ in the monotony of foliage. Fortunately, the leaves often have a coal crust, a phytolemma, from which it is usually possible to obtain a cuticle preparation. The cellular structure of the leaf, the stomata and the hairs do not vary so much from one sheet to the next. Thanks to all these microscopic features, the history of ginkgo is now more or less uncovered.

EARLY BURIED

Do not think that biologists and botanists have already completed an inventory of animals and plants that now inhabit our planet. Every year, dozens of articles appear in the literature describing new species and even genera, whose existence was not suspected before. Very interesting and unexpected, for example, was the discovery in 1957 in Central Peru of a new species of Plouuniform Stylites, a possible descendant of tree-like giants of the Carboniferous period. Usually, botanists and zoologists do discover the genera and species for science, but sometimes after a careful study of the find it turns out that something similar has already been encountered by paleontologists.

In 1941, the Japanese paleobotanik Shigeru Miki found in the upper Tertiary deposits of his country a kind of twigs of coniferous, like a sequoia. A detailed study of the remains showed that this is a new genus. Miki gave him the name metasequoia. Later the same remains were found in the Tertiary and even Cretaceous sediments almost throughout the Northern Hemisphere, up to Spitsbergen. It turned out that many fossil leaves, which used to be used for yew or sequoia, actually belong to metasequoia. Gradually, materials on this plant accumulated, and suddenly in 1944, Chinese botanists found him alive and healthy in Central China, Metasequoia live here on an area of only 800 square meters. Km at an altitude of 700 to 1400 m. Several specimens donated to Soviet botanists can be seen in our botanical gardens. In metasequoia, soft and flat needles, sitting on thin branches in two rows (Figure 42). Finally, the paleobotanists could look at the whole plant, instead of adding it in parts by parts from the disparate remnants. To the credit of paleobotanists, it must be said that they understood well these remains. In any case, the live metasequoia, with all her interest in it, did not force them to radically break their views (Graham, 1999).

Figure 42: Metasequoia (shoot with leaves and cones).

It is rather difficult to answer this question. First of all, it is necessary to clearly define what "living fossils" are. After all, all the living beings surrounding us did not appear yesterday, but have a more or less long geological history. Therefore, they are all living fossils to some extent. Therefore, we will put the question somewhat differently, namely: which of our contemporaries have the most respectable history? The answer will probably be quite unexpected for many. This is not ginkgo, and not metasequoia and not some original exotic plant. It is possible that the most ancient of modern plants are herbaceous plawn-shaped Selaginella and Lycopodium. The first of these plants, the Selaginella, has a creeping or erect shoot, the size of a few centimeters to a good two dozen meters. Although in the modern forests of the Selaginelli there are no close relatives (although, apparently, in the geological past it was "pretty lonely"), one can not say that this kind of genus felt bad. It covers over seven hundred species and, preferring tropical forests, also feels well in temperate climates. In addition, one species has taken root even in the deserts of Mexico and the southwest of the United States (Graham, 1999).

A noteworthy feature of Selaginelli is a variety of cones. This means that in one bump are collected sporangia of two kinds. Some are filled with a mass of microspores with a diameter of 20-70 microns, which sprout even in sporangia, and when they are sown, they give rise to a small sprouting with male genital organs. Other sporangia include 8 to 40 megaspores. They also germinate directly in sporangia, and then in the ground they give rise to female sprouts. A notable feature of Selaginelli is also its foliage. There are two types of leaves on the shoots: one is tightly pressed to the stem, while the others are spread out in two rows in the plane of the stem (Figure 43).

Figure 43: Selaginella from Carboniferous sediments; On the right - modern Selaginella.

Exactly the same remains from time to time are found in the Cenozoic, Mesozoic and even Paleozoic deposits of different countries. The most ancient find was made by the American paleobotanist W. Darr in the coal kidneys of the Carboniferous deposits of the State of Illinois in the USA. The age of these remains is almost 300 million years. Nevertheless, their safety is exceptionally good. In a small oblong cone with a diameter of only 3.5 mm, sporangia with megaspores (four in each) are clearly visible. The spores have already begun to germinate, and the fossilized nuclei are seen in the cells. Moreover, in some cells some formations are visible, which Darra recognized as chromosomes. Over the long history of the Selaginelli appeared features of high specialization in the anatomical structure, for example, real vessels. They were discussed in Chapter IX.

The typical forest plains (Figures 44a, b) resembles the Selaginella in appearance, but, first, leaves of one sort and, secondly, the spores in sporangia are the same. Plains live now in many countries in both tropical and temperate climates. In the tropics they often settle on trees, and in our forests they spread along the earth. The range of economic use of the modern planet is very peculiar. His shoots are sold in cemeteries to decorate graves, they are also laid for the winter for beauty between the windows' frames. Plain disputes are used as a baby powder and, most importantly, in the foundry industry (Marbà, et al., 2007). They sprinkle molds so that metal parts do not stick to the walls.

Figure 44 a: The modern placenta (Lycopodium).

When shoots are found in rocks, outwardly similar to the plains, they are usually called not Lycopodium (ie, placenta), but Lycopodites. The suffix "ites" is often added to the name of the modern genus when it comes to fossil remnants. The very name Lycopodites was unlucky, as it was not always used for its intended purpose. In the 30's, our paleobotanist, MD Zalessky, found in the Upper Permian sediments thin shoots with delicate leaves, which he considered as coniferous and named Walchia spinulifolia. Another paleobotanist, G. P. Radchenko, determined this species in the Upper Permian deposits of the Tunguska basin. It seemed dubious to him that this was really a wahia, and after the name of the family, he just put the question mark in case. After a while, Radchenko felt that the remains were not coniferous, but placenta, and called them Lycopodites spinulifolius. Moreover, he called all the accompanying flora "lycopodite." But then there was an article by MF Neiburg. Not trusting the external resemblance, she applied microscopic technique. As a result, it turned out that these are not mushrooms, but mosses. Radchenko had to rename the flora in the "glottophyll" by the name of another plant. It is possible that some of the other Lycopodites described in the literature are illegally considered to be flat.

Figure 454 b: Plune Lycopodites from Carboniferous Deposits.

Nevertheless, this genus is indeed very ancient. Its remains have been described by German paleobotanists R. Kreizel and G. Veiland from the Upper Devonian deposits of Eiffel (FRG). Their age is not less than 350 million years. From the sporangia it was possible to extract rather large (90-120 microns), but identical spores. The remains of plants, very similar to Lycopodium, are good enough for preservation, they are also found in Carboniferous sediments. Some paleobotanists believe that the oldest of the planetoid - the barraguation from the lower Devonian of Australia (almost 400 million years) is closely related to our planet.

In the popular scientific literature, it is sometimes said of the planet as a herbaceous descendant of treelike coal-bearing lepidodendrons. Observations of recent decades show that this is not so. The placenta is not younger, but, perhaps, even older than lepidodendrons. Most likely, once in the very early stages of the evolution of the well-known among them, two large independent branches have emerged. One of them, although flourishing in the Carboniferous, has now almost completely died out. To her descendants, perhaps, only modern half-faces and the Peruvian style mentioned above are relevant. The other branch has never played an outstanding role, but has successfully survived to the present day. In the organic world, events often developed according to the principle "there is no great Patroclus, the contemptible Tersit lives. (Marbà, et al., 2007)"

The next plant, which to the same extent as ginkgo, has the right to be called a living fossil, is the well-known horsetail (Figurc 45), in Latin Equisetum. Horsetails are quite small in our country, but in the tropics some species reach two meters in height with a diameter of 10 cm. A South American type of horsetail, nicknamed "giant" (Equisetumgiganteum), grows to 12 m in length, but its stem reaches only 3 cm. It is a horsetail climbing along neighboring trees. In the modern flora, only thirty species of this genus, mostly inhabiting the temperate belt of the Northern Hemisphere, are counted.

Figure 46: Modern equisetum horsetail.

The most ancient remnants of real horsetails are found in the Upper Triassic sediments. Separate finds of shoots resembling horsetail are also found in older strata, up to the Carboniferous, but their definition should be further clarified. Triassic and, especially, Jurassic horsetails, in essence, are no different from modern ones. They had the same articulate stems, small fused leaves and cones on the top of the stem. The stems inside are hollow, and in the nodes they are blocked by round diaphragms. When the plant died off, its stem was easily broken and the diaphragms fell out. We often find these neat circles in the Mesozoic and Cenozoic rocks.

Chapter II already said that the needles have an interesting mechanism for spreading spores. From each spore there are narrow ribbons that are twisted in the form of a spring, until the spore is completely ripe. Then, when the mature sporangia is cracked, these springs straighten, the spores pop up and be easily picked up by the wind. Paleobotany repeatedly gutted the sporangia of fossilized horsetails and found in them spores, but only without springs. This puzzles them a little and raises doubts as to whether they really deal with fossil horsetails. However, it is quite possible that under the influence of chemicals, which are subject to spores during their study, the springs simply dissolve (DiMichele, 1996). Let us note, by the way, that the spores with springs were recently discovered by the Americans R. Baxter and D. Leisman in a sporiferous cone of calamite in coal coal of the Carboniferous age (Fig. 46). They dissolved a small piece of rock with the spores inside it in dilute hydrochloric acid. While the spores were in acid, their springs remained folded. But when they were transferred to 50% alcohol, the springs came to life. "This process," wrote Baxter and Leismann, "could be observed under a dissecting microscope, and it was an extraordinary sight to see spores turning under the influence of the rapid movement of the elater (after springing - SM) after so many millions of years inactivity ".

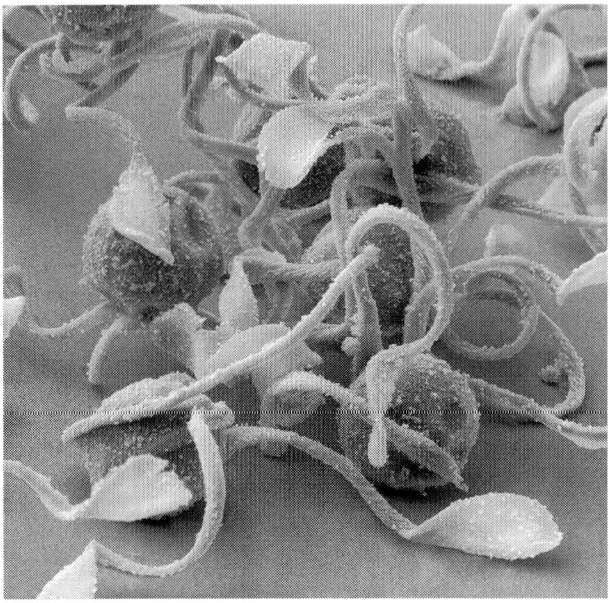

Figure 47: Spores with elastics in modern horsetail.

List of living fossils, undeservedly forgotten by popularizers, can continue for a long time. These are ferns (Osmunda) and Marattia (ferns), which also lived together with dinosaurs. This also includes some coniferous Cretaceous and Jurassic periods. The fact that they belong to the modern genera Podocarpus, Cephalotaxus, Athrotaxis, etc., is shown by the Swedish paleobotanist R. Florin and our paleobotanist VA Krasilov. In any case, these plants are much older than such a classic "living fossil" as metasequoia (DiMichele, 1996).

AGAIN, THE LESSONS OF CAUTION

If the palaeobotanists of the first half of the last century were asked to write living fossils known to them, the result would be a kind of list. There would not have been some of the ones just examined - and at the same time, there would have been other genera. True, even then paleobotanists were rather cautious and avoided applying to the fossil remains the unchanged generic names of modern plants. They widely used the suffix "ites." So entered the literature, for example, calamities - in Latin Calamites from the word Calamus, ie reed (also called modern palm-rattan). It was awkward, of course. After all, calamites and reeds are not relatives at all. A reed is a monocotyledonous (i.e., flowering) plant, and the calamites belong to a very different group of plants (to arthropods). This error was revealed very early and was clear in the last century. Nevertheless, Adiantites (Adiantum, or "Venus of Hair" - fern, and the name Adiantites received a seed fern), Yuccites (in honor of the lily yucca some gymnosperm was named), Zamites (from Zamia), Casuarinites (from Casuarina), etc. In addition, in the lists of Mesozoic and Paleozoic plants ferns of Asplenium (the stone), Cyathea (ciatea) and other modern plants flashed. Each time the choice of the name paleobotanyku hypnotized external similarity. Of course, he knew of the numerous cases of simulation of forms in plants not related by kinship, and yet the similarity bribed. Although blatant mistakes were revealed all the time, the lessons did not go on for future use. Even now, when a serious theoretical basis is brought under the phenomena of parallelism, paleobotanists often continue to believe this notorious external similarity.

Careful paleobotanists, finding the leaves familiar to modern plants, do not hurry to refer them to still living genera and replenish lists of living fossils. It is not enough to find one similar organ: a leaf, a stem or a bump. We still have to figure out whose sheet it is, on which stalk he sat with and with which cones. It's not just that different plants have the same leaves or pollen. The fact is that the evolutionary development of individual organs in plants proceeds unevenly. This phenomenon AL Takhtadzhyan proposed to be called heterobathmium, that is, different stages. First, the shape of the leaf and the structure of the pollen begins to progress, for example, the structure of the cones is tightened, and the anatomical structure of the trunk lags (DiMichele, 1996). Then she starts to make up for lost time. Only when all the organs develop to the required level can we say: "This is a pine (oak, cabbage, etc.)."

The problem of living fossils is of great scientific importance. Considering the question of evolutionary progress (more about him in the next chapter), the largest Darwinist T. Huxley drew attention to the fact that many organisms, remaining unchanged, existed from the Paleozoic to our days. Hence, he concluded that "an amazingly small progress" in the organic world during the past geological epochs.

The above facts about living fossils would probably please Huxley. However, they hardly confirm his conclusions. Not anticipating the forthcoming arguments about progress, let us say only that most of the listed organisms, though playing a prominent role in the landscapes of the geological past, did not play this role all the time. Ginkgoals have died out almost without a trace, selaginelli, horsetails and plaunas have never been the main background of vegetation. The same can be said about other plants with which we have become acquainted in this chapter.

REFERENCES

1. DiMichele, W. A., & Bateman, R. M. (1996). Plant paleoecology and evolutionary inference: two examples from the Paleozoic. *Review of Palaeobotany and Palynology*, *90*, 223–247. https://doi.org/10.1016/0034-6667(95)00085-2
2. Marbà, N., Duarte, C. M., & Agustí, S. (2007). Allometric scaling of plant life history. *Proceedings of the National Academy of Sciences of the United States of America*, *104*(40), 15777–80. https://doi.org/10.1073/pnas.0703476104
3. Graham, A. (1999). Studies in neotropical paleobotany. XIII. An oligo-miocene palynoflora from Simojovel (Chiapas, Mexico). *American Journal of Botany*, *86*(1), 17–31. https://doi.org/10.2307/2656951
4. Kenrick, P., & Crane, P. R. (1997). The origin and early evolution of plants on land. *Nature*, *389*, 33–39. https://doi.org/10.1038/37918

12 WHO IS IN CHARGE OF THE PARADE?

Any modern objectively written course on this or that theoretical section of biology is more like a task book than a set of firmly established provisions. We can list what we are given, but when we have to answer what is asked, most of the time we are at a dead end. In this amazing task book, nature forgot to insert a sheet with answers. And the more important the question, the further we are from his decision (Pfefferkorn, et al., 2014). In the words of P. Erlich and R. Holm, "the basic question remains unanswered: why during the evolution of DNA has created for their own reproduction of tubalons and humans, while bacteria and other simple organisms seemingly can serve no worse for this purpose ? ." We will not discuss this eternal topic, in which there is something akin to the question of the meaning of life. Remi Shovsp rightly writes: "From the fact that we will ask ourselves such unsolvable questions, science, really, will not gain anything." And so we will try to ask less "why", but we will pay more attention to the question "how?".

WHO SHOULD ANSWER THE QUESTIONS?

How did the groups of plants change over time? What events determined these shifts! Similar questions researchers have been asking themselves for a long time. It would seem, who, if not the paleobotanists, should answer them? After all, at their disposal are the relevant "documents." Indeed, the information available to paleobotany is enormous. Following from the depths of the geological past to our days, we see that the era of the rule of ancient psilophytes, prairat and other spore plants was replaced by the era of gymnosperms. Angiosperms came to the fore in the last stage of plant history. And yet paleobotanical documents are largely silent. Not only do fossil plants reach us in disassembled form. The incompleteness of the geological record is manifested in the other (Pfefferkorn, et al., 2014).

We have already told you that a change in the flora of various types occurs somehow suddenly. At certain moments in history it is not clear where the new groups of plants come from, and at once in large numbers. Then they suddenly disappear, leaving a few surviving offspring. So it was with the devil's filial flora, so the Mesozoic flora with its ginkgo, cicadas, Czekanowski and Bennettites appeared and disappeared, so the angiosperms that surrounded us were on the scene. The sudden appearance of new plants and an entire flora in the geological section is not evidence of their occurrence at this time. They lived before (for a long time - it is unknown), and we see only traces of their rapid resettlement. To penetrate into the smithy of new forms of paleobotany still can not, except in exceptional cases. But even then one can not confidently talk about the specific cause of form formation.

Therefore, many (albeit far from all) questions concerning the mechanism of the evolution of the vegetation cover and the plants themselves have to be answered by botany. Of course, only paleobotanical observations can show the actual course of events.

THE BACKGROUND OR THE DRIVING FORCE?

At the cradle of each plant are evil and good wizards. Some prophesy of life's adversity, others wear properties that allow them to defeat their enemies. The plant spends its entire life in the struggle for water, light, space, in the fight against diseases, parasites and herbivores. From the plant's point of view, the cow is a ruthless predator (Weiss-Krejci, 2001).

The importance of the external environment in the life of plants can not be depreciated. Therefore, the sympathies of the majority of researchers to the teaching of Darwin, who believed that it is necessary to see the driving force of evolutionary development in the external environment, is quite understandable. The most important argument in favor of this concept is the adaptability of organisms to various environmental conditions. However, in science questions are not resolved by a majority vote. Many biologists regard the teaching of Darwin with a fair amount of skepticism, and this is not due to their bad character, but to the fact that this doctrine does not answer all the questions convincingly enough. In the words of E. Sinnota, "in the last decade of the 19th century, the enthusiasm caused by the idea of evolution gave way to a more sensible idea that this theory does not resolve all biological problems." There are many such problems, but we will linger only on those of them with which the paleobotanic deals. Just do not think that the author wants to question the essence of evolutionary teaching. It is, on the contrary, a possible explanations from his position of certain facts from the history of the plant world (Pfefferkorn, et al., 2014).

In the tenth chapter, we briefly touched on the final phase in the development of terrestrial plants. Approximately one and a half to two million years ago glaciers began to attack the vast territories of the Earth. Preparations for glaciation, apparently, began quite a long time, somewhere in the depths of the Mesozoic, when after the late Triassic "pantropical" climate, our climate became increasingly sharper and sharper in manifesting its climate differentiation. And yet, apparently, the glaciation erupted quite suddenly. It was a great event in the life of terrestrial plants. It is natural to expect that such a drastic change in conditions should also cause a corresponding evolutionary outburst, but no traces of any outbreak can be seen. In the fourth period and the preceding 10-15 million years, we can not note the appearance of any large groups of plants, for example, orders or families. Moreover, in the Neogene sediments there are almost no genera unknown in the modern flora, although the geographical distribution of all genera was then different than now. Thus, the Neogene and Quaternary history of the plant world seems to give a serious argument to the disposition of the opponents of Darwinism. Where is it here, is it the influence of the external environment?

The next question is given by the Upper Permian flora of Western Europe. We have already said that the entire Permian history of the West European flora was under the sign of a progressive drying up of the climate. It is not necessary to prove that the greater or lesser dryness of the climate is a very important factor of the external environment, to which the evolution of plants must respond. In fact, the Upperperm flora of Western Europe is composed of plants that are also known in older sediments, but play a subordinate role there. This fact seems to please the supporters of the "preadaptation theory."

This theory is based on the notion of natural selection, which plays the role of not a constructive factor, but a kind of filter through which only those plants (or animals) penetrate the territory with changed external conditions, which have already been adapted (adapted) to them For one reason or another. In other words, only shade-tolerant plants penetrate the canopy of the newly formed spruce forest, only those plants that do not die from excess salt will be retained on the become saline soil (Weiss-Krejci, 2001). Hence the conclusion that it is not the environment that creates the signs and habits of organisms, but organisms look for an environment that matches their characteristics and habits. Something similar happens to doctors in a country that has long been free of smallpox if an epidemic suddenly bursts out. Rare and, it seems, unnecessary experts are in the center of attention, become irreplaceable. These arguments can not be denied logically, and the analysis of the fossil flora of arid regions seems to confirm the theory of preadaptation (Pfefferkorn, et al., 2014).

This does not exhaust the "anti-Darwinian" paleobotanical documents. The sources of numerous cases of parallelism in unrelated plants that live far from each other and under quite different conditions are largely unclear. This is the similarity of the Gondwana Buriadia and the coniferous Northern Hemisphere. It is equally significant that leaves of the shadowioptery type appeared in a variety of groups, not related to each other by time, place, or (as if) the conditions of life.

Attention is drawn to the uneven development of different groups under the same conditions, in the same place. Some groups progress, give many species, while others live on the same background for a long time, dispensing with significant innovations. Tricky questions are asked and the most ancient plants. We have already spoken about their sudden appearance and unusually rapid progress, although there were apparently no outstanding stimulating events in the external environment at that time. Finally, the sequence of folding and mountain building phases that changed the face of continents does not coincide with the most important stages in the evolution of terrestrial plants (Weiss-Krejci, 2001).

How can one interpret these and many other paleobotanical facts from the standpoint of Darwinism? Is it really that the external environment is only the background on which evolutionary processes unfold?

Why "or ... or ..." ?

Before answering the questions posed, we clearly distinguish two concepts: the origin of organisms and their appearance in the geological context. Both concepts are not so closely related to each other, as it seems at first glance. In considering general problems of evolution, paleontologists like to write off the lack of documents for the incompleteness of the geological record. This incompleteness is really so significant that the famous German botanist K. Gebel compared the paleobotanical evidence with the pathetic scraps of a distracted library. However, there are no grounds for such pessimism. The loss of pages from the geological record is a natural process, and their absence is eloquent enough by itself (Weiss-Krejci, 2001).

Here is one example. Studying the patterns of distribution of cultivated plants, NI Vavilov drew attention to the fact that on the Earth globe there are places with a huge variety of varieties of this or that plant. Thus, in Ethiopia no more than half a million hectares of land were found, more species and varieties of wheat were found than in all other parts of the globe combined. The works of Vavilov and his students showed that such concentrations are not accidental. They point us to the centers of formation, some generators of new species and varieties. Moreover, Vavilov writes about the facts "of the amazing confluence of the process of basic formation of geographically to extremely small spaces." It turned out that the modern centers of high-quality riches are confined to the mountainous regions of Asia, the Himalayas and their spurs, to the mountain systems of North-Eastern Africa, southern Europe (the Pyrenees, the Apennines, the Balkans), the Cordilleras, the southern spurs of the Rocky Mountains (Fig. 47). In the Old World it is a question of a strip between 20 and 40 ° with. W. These mountain areas border on the deserts of Central Asia, the Sahara and on the diversity of climate and soils represent the optimal conditions for the manifestation of the formative process. By the amount of precipitation, by temperature differences, by soil types, gradations to the extreme variants are expressed here. It is also important that "the geographical location of the speciation of cultivated plants largely coincides with the localization of the general speciation process characteristic of the floras of the entire Earth."

From what has been said, a very important conclusion for paleobotany follows. Since the vast majority of ancient plants, known to paleobotanists, inhabited swamps and lowlands, and not foothills at all, there is little chance of meeting "documents" about form formation in the geological past. A different conclusion follows from this: a change in the flora, often a very sharp one, which we see in a section, has no direct relationship to speciation. This is the next stage in the development of the vegetation cover, namely, the resettlement into new conditions of the flora, formed behind the scenes of the events observed by us.

Charles Darwin, the English paleobotanist DG Scott (in the 1920s) wrote about this mechanism of vegetation development in the geological past, but it was most fully revealed in the works of our largest paleobotanist A. N. Krishtofovich. The behavior of plants with a sudden change in external conditions (glaciation, aridization), AN Krishtofovich compared with the panic flight of a cat climbing a tree at the sight of a dog. Indeed, plants that are well adapted to the external environment need time for rearrangements, if this environment changes. Rapid changes in climate and soil cause the plant to die. For the evolution, homeopathic doses of directed changes are needed (Lin, 2010).

The notion that the main form-building takes place in elevated areas of land and therefore falls out of our field of vision was widely adopted by paleobotanists. Now this phenomenon is referred to every time they see the sudden appearance of groups of plants and a whole flora. But what about the early Devonian events in the world of terrestrial plants? Is it even here that a sudden appearance in the context of a rich fossil flora depends on the same reason? No, hardly. It is very difficult, almost impossible to abandon the notion that the cradle of terrestrial plants is the coast of seas or other water bodies. The ancestors of terrestrial plants were algae. This is the unanimous opinion of the researchers, and there is no reason to question it. Assuming that the algae descended into the lowlands, that is, into the region of formation of the floriferous strata, having begun its journey in the foothills, at least boldly. The reason for the sudden appearance of the wealth of the Early Devonian flora, apparently, must be sought in another (Lin, 2010).

Vegetation is not always covered continents. Once upon a time the land was completely lifeless. The first plants began to penetrate into the depths of it through separate moistened areas. In these conditions, there was no real soil formation on the continents, nor was the flow of water regulated. It is possible that, for this reason, all continental and coastal-marine sedimentation was not at all the same as in the last geological epochs. In other words, the "cemeteries" of the oldest plants were not sufficiently stable either. As the evolution of terrestrial plants became capable of a more complete settlement of the land, the development of the root system allowed them to be more firmly retained on the substrate, and hence to anchor it.

Even a small acquisition in this respect could dramatically change the whole situation. A single connected chain is formed: more plants, they sit more tightly - drain is better regulated - less erosion of slopes - less damage in the plant cover - more plants, etc. The process has increased in avalanche, in a very short time. The land quickly acquired a lived-out appearance, quite tolerable "cemeteries" of plants began to be formed, but the coal industry had not yet begun. Later the same processes, developing and deepening, changed the character of the runoff of water and soil formation so much that it became possible to form marshy lowlands in which coals began to accumulate. By the way, the latter appear in the geological section also very unexpectedly. Devonian coals are rare, but in the Lower Carboniferous sediments they are already abundant in many parts of the globe. Expansion to unoccupied areas of land with little competition and a rich influence of factors, each of which was a wonder to the descendants of water, is, apparently, the source that the Devonian plant world owes its ultra-rapid evolution (Lin, 2010).

On several tricky paleobotanical questions, we seem to have picked up more or less tolerable answers. One can also explain the unequal rates of evolution of different plants under the same conditions: first, the changes do not necessarily capture all components of the environment in equal measure; Secondly, different plants have not the same measure of evolutionary plasticity.

It is more difficult to explain from the point of view of the Darwinian theory of natural selection numerous cases of parallelisms, of which much has been said in the ninth chapter. In any case, complex genetic mechanisms must play a huge role here. It is impossible to exclude and possible manifestations of directed development occurring under a certain autonomy from the external environment (Lin, 2010).

The question of the origin of parallelisms can not be finally solved, until the adaptive value of the parallel appearing signs is elucidated. Before that, it is still too far away, and we are forced to occupy rather a conciliatory position with respect to the influence of the environment, without putting the question "or ... or ..." firmly, remembering that most of the cardinal issues of science found a solution when the researchers accepted Into account "and ..., and ...." Similarly, the whole concept of preadaptation can not be completely rejected. In a polemic with her supporters, the paleontologist Pompetsky rightly remarked that the ichthyosaurs had hardly passed to life in the sea, for the reason that their limbs began to transform into fins (Lin, 2010).

However, when we are dealing with rapid and strong changes in the environment, we see that only those who can live in these extreme conditions penetrate these areas from the surrounding areas. Here there is not so much a change as a migration of organisms. For example, in the tertiary (especially early Tertiary, i.e., Paleogene) time on Earth, apparently, there were no wide areas with an arctic climate. At that time, the tundra and forest tundra were obviously such an extreme and atypical environment. But the Quaternary glaciation dramatically changed the picture. And today the cold climate prevails in vast territories. The latter were populated, apparently, mainly by those plants that already learned how to do without much heat.

IS PROGRESS ACCELERATING?

Among the main problems of biology, to the solution of which paleobotany (and, in general, paleontology) has a direct relationship, there is the problem of evolutionary progress. Among foreign scientists there are many who deny organic progress or consider it insignificant. Even the famous Darwinian Thomas Huxley, a passionate preacher of evolutionary teaching, wrote that the "total result" of the development of the organic world "is amazingly insignificant." This point of view can be contrasted with the statement of paleobotanist G. P. Radchenko, who categorically categorically writes: "The progressive evolution of the organic world is expressed in raising the level of organization of animals and plants throughout the history of the biosphere and has a clearly expressed tendency towards a gradual acceleration of this process." But one of the greatest modern geneticists, the Swedish scientist A. Myuntzing writes: "... At present, the process of evolution towards increasing complexity of the organization is hardly going on." We will confine ourselves to these atrocious utterances, since it is completely impossible to bring all the points of view laid out in the literature. Considerable scientific articles and entire monographs are devoted to their consideration.

The question of evolutionary progress for all its seeming simplicity (is it really the development from amoeba to man not a sign of progress?) Can not be considered definitively resolved. Suffice it to say that genetic mechanisms are surprisingly similar in the vast majority of organisms. Therefore, genetics, studying the heredity of fruit fly, can apply many of the resulting patterns to humans. Thus, for such an important feature of all living things as heredity, we are not able to indicate any significant progress during the last half-billion years.

The difficulty in determining progress is also connected with another circumstance. Giant dinosaurs, if they were endowed with reason and could get acquainted with the remains of their ancestors, would probably consider themselves sufficiently progressive. Nevertheless, dinosaurs, once dominant on land, sea and air, died out, and their distant primitive ancestor cystepera fish latimerium survived to the present day. Assessing progress in terms of the height and complexity of the body is a risky business (Lin, 2010).

Often the progressivity of the animal is judged by whether it was on the phylogenetic road leading to man. We ourselves have become in the eyes of many biologists a measure of progress. This is, of course, a rather shaky criterion, because the fate of mankind on the scales of geological time is unknown. Who knows, maybe Remi Shaven will be right. He jokingly writes that "years from now a million in a hundred some six-legged erudite will write that judging by very ancient sources, a certain monkey has suddenly turned into another being and this new being has tuned the horror of what, some mountains from Stone, and then immediately disappeared ... " (Lin, 2010)

Nevertheless, it is difficult to completely deny progress. Following most of our researchers, we assume that progress is at least very likely. If so, it is interesting to determine the extent of the progress of plants in different geological periods. Here, obviously, you need to evaluate by two indicators: first by the appearance of certain features, and then by their combinations. And here we come across very interesting facts. We divide the terrestrial plants known to us into individual features and see in what order they appeared in the geological history. It turns out that even in the early Devonian, the main features of the internal structure were outlined. At the end of the Devonian internal structure has already reached a very high differentiation. From this time to the Mesozoic, there are no significant improvements in it. True, in the second half of the Mesozoic, angiosperms appeared, whose important acquisition was the vessels. Devon plants have succeeded not only in the anatomical structure. They already had ovules. Until the appearance of angiosperms, the seeds did not undergo significant complications. The angiosperms of the angiosperms are covered with fused coverlets, and the process of fertilization has become more complicated.

As for the structure of the foliage, here again the main successes also belong to the Devonian plants. True, they still do not have leaves, in which the veins are located in the cirrus, and also leaves with a reticular venation. However, all this appeared already in the middle of the Carboniferous period (Huston, & Smith, 1987).

A great achievement was the acquisition of angiosperm angiosperms. The advantages of herbaceous (especially annual) plants have already been discussed above. When such angiosperms appeared on Earth, we do not really know, but it is possible that they were already in the second half of the Cretaceous.

Unfortunately, we can not make the same evaluation of individual traits, but of whole organisms, because of the incompleteness of those plant remains that fall into the hands of paleobotanists. However, and said enough to suggest that in the history of terrestrial plants, the main progress was made in two stages. The first stage falls on the Devonian period, the second - in the middle of the Mesozoic. In recent geological epochs, plants, engaged in small-scale improvement, have not achieved anything unusual.

Thus, from the three points of view on the progress, named at the beginning of this section, one has to lean towards the point of view expressed by the geneticist Myintzing. We can not deny progress in the plant world, at least with today's state of our knowledge. He clearly affected the external appearance of plants and allowed them to seize vast areas. But it is still difficult to talk about any decisive acceleration of plant progress. Rather, on the contrary, now we are witnessing their calm evolution, which does not lead to a significant complication of the organization. This is quite understandable in conditions of land abundantly populated by perfectly adapted organisms, experienced competitors in the struggle for existence.

IF WE FELL IN UNISON ...

The fact that the questions discussed above are interesting, there is no doubt. They are not only interesting, but also important, and we will talk about their meaning in the next chapter. However, they are still far from solving them. However, this is the fate of all the cardinal issues of biology. The decision is delayed by the fact that joint efforts of specialists in different fields of knowledge are needed for the nest: botanists, paleobotanists, geologists, geneticists. As it is now fashionable to say, an integrated approach to solving these problems is needed (Huston, & Smith, 1987).

Indeed, the evolution of fossil plants can not be studied in isolation from modern biology. Many curious biological mechanisms have already been established, the evolutionary significance of which has not really been appreciated. For example, only a few paleobotanists take into account neo-genetic phenomena in their evolutionary constructions. It is about the ability of organisms to start reproducing until the completion of individual development, sometimes even in the embryonic state. In the works of AL Takhtadzhyan, IT Vasilchenko, and other researchers, it is clearly shown how important the neoteny played in the evolution of angiosperms. AL Takhtadzhyan believes that it is with neoteny that the appearance of a herbaceous appearance in many angiosperms, and the very angiosperm, is also associated. We know very little about the phenomena of neoteny in other groups of plants.

Here is another example. In the works devoted to the general problems of paleobotany, you hardly find the word "apomixis." This term denotes cases of the replacement of sexual reproduction by asexual. Such asexual reproduction can be both vegetative (by means of layers, tubers, bulbs, etc.), and with the appearance of externally ordinary seeds, which in fact develop without fertilization. Plants-apomicts have a number of advantages, sometimes decisive. Due to this phenomenon, a species well adapted to these conditions can quickly reproduce in mass quantities with an absolutely identical and, moreover, optimal genetic structure. During sexual intercourse, variability, often harmful, is much greater. Apomicts invariably bring a lot of trouble to taxonomists, because instead of a wide range of variability they form a set of constant types, each of which differs little from each other. To the apomicts, by the way, is the hawk, which gave a somewhat different splitting of the characteristics of the hybrids than the peas, which made G. Mendel doubt the universality of the laws he derived. Apomictic plants are a real punishment for the taxonomist, who, seeing the consistency of the changes (albeit small ones), describes more and more new species, in fact dealing with the same species. Perhaps, taking such phenomena into account by paleobotanists will force them to be more cautious and reduce the number of new types of fossil plants described in the literature (Huston, & Smith, 1987).

The number of similar examples, when genetics can come to help in solving paleobotanical problems, can be easily increased. Experience shows that the union of genetics and paleozoology is very fruitful (remember the most interesting works of JG Simpson). Why is the science of fossil plants an orphan?

Exactly the same close ties can be established by paleobotany with paleogeography, the theory of the accumulation of sedimentary minerals, the geography of plants and many other disciplines. But, unfortunately, these connections are weak and not permanent. This affects both paleobotany and the listed fields of knowledge. What is the matter, what prevents everything from placing in its place, what prevents the establishment of all these fruitful ties? What is it - the fault or trouble of paleobotanists? With this question, we will try to understand the next chapter.

REFERENCES

1. Lin, R. (2010). [Summarization of Chinese rhubarb recorded by the Westerners in China in the Ming and Qing dynasties]. *Zhonghua Yi Shi Za Zhi (Beijing, China : 1980)*, *40*(2), 80–6. Retrieved from http://www.ncbi.nlm.nih.gov/pubmed/21118201
2. Huston, M., & Smith, T. (1987). Plant Succession: Life History and Competition. *The American Naturalist, 130*(2), 168. https://doi.org/10.1086/284704
3. Joyce, R. A. (2004). Tikal: Dynasties, Foreigners & Affairs of State: Advancing Maya Archaeology (Book). *Choice: Current Reviews for Academic Libraries*, *41*(6), 1121.
4. Weiss-Krejci, E. (2001). Restless Corpses: "Secondary Burial" in the Babenberg and Habsburg Dynasties. *Antiquity*, *75*(290), 769–780. https://doi.org/10.1017/S0003598X00089274
5. Pfefferkorn, H. W., Alleman, V., & Iannuzzi, R. (2014). A greenhouse interval between icehouse times: Climate change, long-distance plant dispersal, and plate motion in the Mississippian (late Visean-earliest Serpukhovian) of Gondwana. *Gondwana Research*. https://doi.org/10.1016/j.gr.2013.08.022

13 DO YOU NEED A LOT OF PALEOBOTANY FOR HAPPINESS?

Conclusions in paleobotany, like any other science, have theoretical and practical significance. First about the first. Probably, now there is no person who denies the importance of evolutionary theory and its influence on the overall worldview. It is quite obvious that paleontology in the formulation and solution of evolutionary questions does not belong to the last role, and that these paleobotanists deserve no less attention than the data of paleozoology. In some respects, fossil plants give even more interesting material for generalizations than fossil animals. After all, only the skeleton remains from the latter, and we often see the internal microscopic structure of the remains of plants in its entirety.

Already this one value of fossil plants is enough to consider their study justified. However, it's not just that. In our country the main paleobotanical cells are organized in geological institutions. Abroad paleobotanical studies are also financed not by philosophical societies, but mainly by geological departments, oil and coal companies. At the beginning of the book it was said that without paleobotany people would not have such coal basins as Ruhr or Kuzbass.

No useful mineral can be developed until its deposit has been studied by all the rules of science. The formation of the deposit is not an accidental, but a natural process. To know the pattern is to analyze the process in space and time. The geologist needs a geological clock, and they are replaced by the remains of plants and animals. In the continental deposits always come to the forefront of plants (Wolfe, 1987).

Of course, the remains of fossil organisms are not the only indicator of the age of the rocks. Moreover, they give us only a sequence of events, and not their duration in millions of years, which is determined by physico-chemical methods. Here the question is pertinent: is it worth to contact with shells and leaves, if you can directly establish the age of rocks physico-chemical way? How things will be in the future (so far distant), we do not know, but now it definitely stands. Methods for measuring the absolute age of rocks have made impressive successes in recent decades, but so far they have not been able to replace paleontology in any way. There are several reasons for this. First, physico-chemical methods are expensive and time-consuming. Secondly, they are far from everywhere applicable. Thirdly, they are still very inaccurate. This can be seen from an example. Another 10 years ago it was believed that the age of the boundary between the Carboniferous and Permian periods was 190-220 million years. Now this border has been pushed aside for more than 50 million years. The position of rocks in the scale of geological periods is determined by paleontological methods several times more accurately than with the help of physics and chemistry. Here is a concrete example. The duration of the Carboniferous period is now determined in the interval of 55-75 million years (the difference in the evaluation of the duration by different observations is 20 million years). The error of each measurement of the age of the rock belonging to this period is not less than 10 million years. According to fossil animals and plants (in those places where they are well studied), the Carboniferous period is divided into 14-15 parts, each of which is recognized quite well. Here, the accuracy of the paleontological method in determining the sequence of events is three to four times higher than the physico-chemical one. Of course, the technique of determining absolute age is continuously being improved, but paleontology also does not stand still. In short, paleontology, including paleobotany, is soon to be archived.

It can not be said that geologists do not know all this. Quite the contrary! They often send samples for analysis to paleobotanists than in an absolute age laboratory. We will not give absolute numbers that will not tell anyone anything. It is better to turn to comparisons.

Among the largest coal basins of our country is the Tunguska. Here, huge reserves of coal (not less than 1500 billion tons) and many other minerals are concentrated. Recently, oil was found in the Tunguska basin. Interest in the geology of the basin is very large, and there are numerous research, survey and exploration parties in it. In the center of their attention - deposits with numerous plant remains. In recent years, collections have been collected here that have filled hundreds of boxes. It's thousands and thousands of prints, often surprisingly secure. But we do not have a paleobotany, who would systematically master this wealth. Collections are sent to different institutions, where they are studied in the very first approximation and fall on the shelves with a dead load. The geologist receives only a small fraction of the information that these plants could give. The area of the Tunguska basin is five times larger than a country like the FRG. In the latter, in addition to Ruhr and Saar, there are no large coal basins with Paleozoic plants. Reserves of coal in the Ruhr and Saar are ten times less than in the Tunguska basin. Nevertheless, Paleozoic plants of West Germany are now studied by at least 10 people. It should be taken into account that these are the same plants as in France, England and other countries of Western Europe. There they also studied and studied (and for 150 years) a large number of highly qualified specialists. Many of them during the whole life were engaged in the plants of one small pool and figured out them in detail. In the Tunguska basin, for a hundred years of its study, no more than ten paleobotanists worked. Half of them were engaged in this pool only in passing with other studies, and the other half - only a few years. The result of this was that on geological maps for many years Permian coal-bearing deposits showed only two colors. It was not possible to divide these deposits into narrower intervals without the aid of fossil plants. Such cards are difficult to read, and somehow they will have to be altered.

Approximately the same situation developed in the Pechora basin. The Paleozoic plants of the Northeast of the USSR, Taimyr and Kazakhstan are even worse studied. In the Kuznetsk basin they are studied more intensively. But there are still a lot of blank spots here. Suffice it to say that to this day we do not know much of the generic and species species of the Paleozoic plants of the Kuznetsk Basin. It's a sad situation with a thorough botanical study of the Paleozoic plants in Siberia. About their anatomy, almost nothing is known. (In this respect, the Paleozoic plants of Western Europe were much better studied in the first half of the last century.) As a result, an enormous number of original and completely incomprehensible plants, each of which is a pantry of surprises, falls out of evolutionary constructions (Putland, & Goldizen, 2001).

All these interesting and necessary studies of our paleobotany to deal with once. They act as a bibliophile who does not have time to read his books, but only by name puts the book on this or that shelf. There are few such books, the essence of which is visible on the cover. Even fewer such fossil plants, which can be understood without resorting to detailed research using modern technology. Meanwhile, the terms of reference and interests of paleobotanists do not narrow, but expand. The interest of researchers of a very different profile to the general problems of the evolution of the living is continuously growing. New territories are being developed, geological study of our country is expanding, general laws are being discussed in changing the face of the planet, and here the paleobotanik can not stand aside.

But this does not end the troubles of paleobotany. You have to be self-taught. Higher educational institutions do not prepare paleobotany (Bazzaz, 2001). The course of paleobotany is here and there read, but so kooky that only the most elementary information can be included in it. To engage in fossil plants come or botanists who do not know geology and have not seen in the face prints, or geologists who know nothing about plants. Paleobotany has to deal with people who were preparing to become specialists in ore minerals, geomorphology, petroleum geology, etc. Of course, many paleobotanists of the past began their work as mining engineers, and such a prominent paleobotanist as DG Scott was first a specialist in locomotives. However, such exceptions cannot be counted. Modern paleobotany is increasingly differentiated. It requires more and more qualified researchers. Most of the specialists only deal with the flora of a certain period and a small area. There is not enough time for more. Not too far and specialization in certain groups of plants (now paleobotanists usually treat the flora entirely).

Meanwhile, the requirements of modern systematics of plants are becoming more complicated. Once palaeobotanik could do with a magnifying glass and a dissecting needle. Then I had to turn for help to the thin sections and the microscope. If the paleobotanist wants to process the material according to all the rules now, he will have to deal with a large number of reagents and complex optical devices. Detailed study of plants requires enormous effort. The study of one type can take many months. However, even if our palaeobotanist overcame all these difficulties, then one, almost the most important, remains. We must publish our work. Articles with a description of fossil plants we can publish only in two journals ("Botanichesky zhurnal" and "Paleontological Journal").

If paleobotanical generalizations still enjoy some success, then the work on the systematic study and description of individual plants and flora seems useless empiricism. Meanwhile, "if Lamarck and Darwin were not systematists, if the former had not written their Flora of France, and the second of their monograph of the barnacles, then we would not have had the Qi Philosophy of Zoology or the Origin of Species." These words, which Academician LS Berg said in the early 1920s, retain their significance even now. The work on the systematization of the entire diversity of even the modern organic world is still far from complete. What can we say about the fossil flora, when each new collection delivers to the attentive researcher something interesting and hitherto unprecedented?

Considering the state of our paleobotany, we have in mind primarily the work on the study of Paleozoic plants. Studies of later plants are more successful, but mainly because the plants of the Mesozoic and Cenozoic deposits in the USSR and beyond are roughly the same. Therefore, our researchers can fully use the results of the work of foreign scientists. When we are dealing with Paleozoic plants of the ancient continent of Angaraida (about half of our country), we have to rely only on ourselves. Outside the USSR, these plants are found only in Mongolia, in which there are no paleobotanists yet, and Northeast China.

When people talk about the need to study wildlife, they usually do not give preference to animals in front of plants. Everyone is clear that the organic world is one, and one cannot study the laws of its development, neglecting the vegetable kingdom. However, when it comes to the organic world of the geological past, this is often overlooked (Putland, & Goldizen, 2001).

Fortunately, among our paleobotanists, there have always been enthusiasts who, despite all the difficulties, have devoted their lives to studying fossil plants and have made many amazing discoveries. Let us recall the names of such scientists as IF Shmal'gauzen, MD Zalessky, AN Krishtofovich, MF Neiburg, IV Palibin, and VD Prinada. However, you cannot rely only on such enthusiasts. Modern science requires the perfect organization of the work of many researchers.

Fossil plants are a very complex object for study. The experience of the world paleobotanika testifies that it is impossible to know them without laborious research using labor-intensive methods. However, the theoretical and practical conclusions that will be obtained will cost a fair amount of labor expended. Our paleobotanists have already proved that they are able to conduct these fascinating and grandiose studies. The best evidence of this is the progress they have made.

REFERENCES

1. Putland, D. A., & Goldizen, A. W. (2001). Family dynasties in the Tasmanian native hen (Gallinula mortierii). *Behavioral Ecology and Sociobiology*, *51*(1), 26–32. https://doi.org/10.1007/s002650100404
2. Wolfe, J. A. (1987). Paleontological Society Late Cretaceous-Cenozoic History of Deciduousness and the Terminal Cretaceous Event. *Paleobiology*, *13*(2), 215–226. https://doi.org/10.1017/S0094837300008769

14 FOR THE MOST CURIOUS

Absolute age of rocks. When a geologist says that an event occurred in the Permian period (Words *in italics are explained in the same chapter*), this does not mean that he knows for sure how many years have passed since. 10 years ago, we thought that the Permian period, established according to paleontological data, lasted 25-30 million years, starting 190-220 million years ago. Now it is believed that it lasted 45 million years, and its beginning moved back into the depth of history to a figure of 273 ± 10 million years. Nevertheless, this chronology in millions of years, with all the relativity of its numbers is called "absolute." Chronology, which is based on the sequence of events without specifying their prescription in millions of years, is much more constant, but it is called "relative." Once at the Geological Institute of the Academy of Sciences of the USSR in Moscow, at a meeting of the Academic Council, it was a question of the difference between absolute and relative geochronology. A voice from the audience noticed that this was essentially the same thing. Academician AL Yanshin objected to this: "The difference between them is enormous." Relative geochronology is usually accurate, while absolute is not. "This, of course, is a joke. The absolute age of rocks is successfully determined by the degree of disintegration of different radioactive elements and the accuracy of measurements is progressing. Other physicochemical methods are used less frequently. See also: "Geochronological scale" and p. 193.

Anthropogenous period (anthropogen) - see "Quaternary period."

Apomixis. Different ways of reproduction without fertilization. Thanks to them, plants that are well adapted to the environment can quickly produce with minimal variability (often harmful) a large number of offspring, equally well adapted.

Bennettity. Extinct group gymnosperms of the Mesozoic era, in appearance resembling cicadas. They had pinnate leaves with complicated stomata. "Flowers" were both bisexual and dioeciously. Since the structure of the reproductive organs of the Bennettites and the most primitive angiosperms is similar, some paleobotanists made a conclusion on this basis about the origin of angiosperms from the Bennettites. Then studies that are more detailed questioned this point of view.

Valhii. The so-called ancient conifers lived at the end of the Carboniferous and in Permian periods. Their leaves looked like spruce needles, but the structure of the cones and trunks was different. After studying the genus of valhia, the Swedish palaeobotanist R. Florin singled out of him two separate genuses Lebachia and Ernestiodendron. He showed how evolution led from these ancient plants to modern conifers. This is in chapter VIII.

View. If biology could say exactly what a species is, half of its problems would be solved. This topic is devoted to a huge literature. Academician V. L. Komarov, considering the question of the species in a special book, came to such a definition: "A species is a collection of generations derived from a common ancestor and under the influence of the environment and the struggle for the existence of living beings isolated by the selection from the rest of the world; The species is a certain stage in the evolutionary process." Academician NI Vavilov believed that the species is "a separate mobile morpho-physiological system, connected in its genesis with a certain environment and area." There are many interpretations of the species. However, as Charles Darwin once said, "none of the definitions satisfy all naturalists." True, further Darwin writes, "That every naturalist vaguely understands what he understands when speaking of the form." We also understand it vaguely. Let us turn to the worldly example. Take the birch. This is the genus, which has the Latin name Betula. In the middle belt, there are various birches. In some, the leaves on the back are bare, and the ends of the branches are thin, hanging. In other birches, the leaves are covered from the back with down and thin branches do not hang. In addition, closer to the north in the middle lane there is a birch in the form of a shrub with rounded small leaves. All these are different types of birch (birch drooping, birch fluffy and dwarf birch). Similarly, the genera are currant (species - red and black), cabbage (species - cabbage garden, rutabaga, turnip, and turnips), etc. It is important to remember that the genus is divided into species, and not vice versa. In the Latin name, composed, as a rule, of two words, the first word corresponds to the genus, is written with a capital letter, and the second - to the mind, and is written more often with a lowercase letter (Zhang, 2001).

Geological map. Image of the structure of the earth's crust on the map. In this case, different colors or shading are distinguished first by rocks of different ages, which in most cases is determined by fossil remnants of plants and animals. Sometimes the map indicates the change in the composition of rocks. Special lines show dissecting the earth's crust. Looking at the contours on the geological map, the geologist can graphically reconstruct the course of strata, including those containing minerals, beneath the surface of the earth. According to this map, the geologist reads the geological history of the area. Mapping is the basis for the geological study of the area.

Geological section. This is the name of the image on the graph of the sequence of rocks composing the earth's crust in some territory. This graph shows the composition of the rocks, some of their properties, the remains of plants and animals contained in them, and other information. The geological section is compiled by studying outcrops, drilling wells and other methods. Geologic maps are often accompanied by a geological survey. The stratigraphy deals with the compilation and comparison of such sections.

Geochronological scale. The history of the Earth is divided into several lengths of time. The largest stretches are the eras. We live in the Cenozoic era, before it was the Mesozoic and even earlier the Paleozoic era. For more ancient eras, united by one word "Precambrian", there are several names

(Proterozoic, Riphean, Archaean, Aza, etc.). Era is divided into periods. The Paleozoic era contains, from the most ancient, Cambrian, Ordovician, Silurian, Devonian, Carboniferous (Carboniferous) and Permian periods. The Mesozoic era is divided into three periods: the Triassic, the Jurassic and the Cretaceous. In the Cenozoic era, the Tertiary and Quaternary periods were formerly distinguished. Now, instead of the tertiary, two periods are distinguished: Paleogene and Neogene. Quaternary period in honor of a man is sometimes called "anthropogenous." In everyday life, the names of periods are usually shortened and the Cambrian, Devonian, Carboniferous, Permian, Jurassic, Chalk, Neogene, Anthropogen etc. types are told. Each period corresponds to its own thickness of deposited Rocks - the "system." Periods are divided into epochs ("Early Cretaceous era", "late-nonpermic era", "Middle Triassic era"). Epochs are divided into centuries. In the thickness of the rocks, the epochs correspond to the departments, and to the centuries - to the tiers. A brief description of the eras and periods is given below under the corresponding letters. See also "Absolute age of rocks."

Ginkgoed. A group of gymnosperms, of which only ginkgo has survived to this day. About it is told in detail in chapter XI ("Tree for forty ecu"). Representatives of ginkgoids are reliably known, beginning with the end of the Paleozoic, namely, with the lower Permian. The leaves are found in a more similar form in older rocks, but it is possible that they are representatives of some other groups.

Glossopterium flora. This flora inhabited almost all of the Southern Hemisphere, viz. Gondwana, during the Late Paleozoic. The name has received from the most widespread plant "glossopteris" in it. This chapter is devoted to this flora.

Glossopteris. Genus gymnosperms with lanceolate or linguiform leaves intersected by a thick middle vein. From it the lateral veins, connecting with each other, leave, so that a network of veins is formed. Glossopteris - the most characteristic plants of Gondwana, the Paleozoic flora of which is called "glossopteria." See chapter V.

Gymnosperms. Plants, multiplying by seeds, are divided into two groups: gymnosperms and angiosperms (flowering). To understand the difference between them, look at the meaning of the word ovum. Seeds are found in all seed plants, but in angiosperms they are hidden in intergrown coverlets (carpels), while in gymnosperms they are completely open or sit in unclosed cups. Modern gymnosperms are pollinated by the wind, and some fossil representatives may have been pollinated by insects. Among the gymnosperms are pteridosperms (seed ferns), ginkgoids, cicadas, conifers, cordayites, bennettites and some other groups of plants.

Homology. Organs having the same origin, but not necessarily similar in appearance. Homologous can be the bodies performing a variety of work. For example, out of the same primordia, the human hand, seal flap and bat wing are formed and therefore have a common origin.

Gondwana. This word, introduced into science in the last century, is made up of two words: Gonda is a tribe in India; Vana is one of the Indian historical regions. A hypothetical continent once united India, Australia, Antarctica, South Africa, Madagascar and South America. About Gondwana and the Gondwanian flora is spoken in V, VI and VII chapters (Zhang, 2001).

The Devonian Period (Devonian). The fourth period of the Paleozoic era, which began 393 (± 10) million years ago and lasted 50-70 million years. The name was received from the county of Devon (Devonshire) in England. The plants of this period are described in chapter I. The animal world was then abundant and varied. During this time, shellfish spread widely, at the end of the period the first terrestrial vertebrates appeared. They on land already had something to eat. By the way, in the Urals, the word "devon" is widespread and even clubs and restaurants call them. The matter is that the best oil from the Second Baku comes from Devonian deposits.

Diatoms. Microscopic unicellular algae having elegant openwork leaves, consisting of silicon. They live in the seas and fresh water.

Precambrian. This word denotes the oldest deposits and the corresponding geological time, preceding the Paleozoic. The Precambrian ended approximately 570 million years ago.

An adolescent. Spore plants in sporangia are maturing with spores that are sown and, if the situation favors, sprout. Tiny plants, called sprouts, are obtained. They develop male and / or female sexual organs.

Spermatozoa, isolated by male organs, reach the ovum. Fertilization is performed. The fertilized egg starts to divide and gives rise to the plant itself. This is how the ferns, horsetails, plains and other higher spore plants develop. In many plants, spores grow in sporangia. In mosses, male and female organs develop on the basic stems, which are actually called moss. These stems correspond to sprouts. After fertilization, thin legs with sporangia on the ends grow from the eggs, which are located on the apex of the stems.

The Cenozoic era (Cenozoic). The last era of the geological history of the Earth, continuing in our days. See "Geochronological scale."

Calamities. Large tree-like plants, together with horsetails and clino-leaves (sphenophylls), united in a group of arthropods. Стр. 46.

Carboniferous period. Both the period itself and the corresponding deposits are sometimes called "carbon" (from the Latin carbo, carbonis - coal). This - the fifth in a row, i.e., the penultimate period of the Paleozoic era. It began 326-345 million years ago and lasted 55-75 million years. Its name dates back to the last century in Europe, where the corresponding deposits contain huge coal reserves. In fairness, we note that in some other periods, not less, or even more coal, has accumulated. About plants of the Carboniferous period is told in II, III and IV chapters. At the end of the period, the first reptiles appeared, but various amphibians dominated on land. Carboniferous are the oldest insects. In the seas, there lived many diverse animals, the skeletons of which accumulated and gave rise to limestones. It was from the limestones of this age that Moscow built white stone.

Catasia (Catalan). In the middle Ages, China was called Katai. Add to this word "Asia" - and get "Catania." It is the name of an ancient continent, located in the past geological epochs (beginning with the Paleozoic) on the territory of China, Korea and Southeast Asia. The flora that inhabited this continent in the Paleozoic is called "the Catalan." It was similar to the flora described in chapter III.

Keithonia. One of the groups of gymnosperms of the Mesozoic era. They had leaves with a reticular venation, and the oedemas are in the form of ovaries. Still, there were no angiosperms, since the entrance to their ovules, although it was safely hidden, did not completely cure.

Cambrian period (Cambrian). The first period of the Paleozoic era, which began about 570 million years ago and lasted more than 70 million years? In the seas, then lived many various invertebrate animals (corals, mollusks, arthropods, worms, etc.). The land, apparently, was lifeless. In the sediments of this period, strange impressions are found, similar to terrestrial plants. Its name dates from Cambria (the ancient name of Wales).

Whipperfish. Almost exactly an extinct group of fish, whose front fins resembled primitive limbs. It is assumed that it was from these fish that terrestrial vertebrates began. Kisteperye fishes have long been considered extinct, but in the thirties, the coelacanths that survived to this day were caught.

Class. The whole plant world is divided into two groups: lower and higher plants. These groups are divided into types (bacteria, fungi, mushroom-shaped, etc.). Types are divided into classes, classes by orders, orders by families. Next are the genera, species, subspecies, etc.

Klinolists (sphenophylls). Extinct group of arthropods. Their name they received in the form of leaves, having a wedge-shaped outline.

Convergence. External and usually superficial similarity of organs not related to common origin, or whole organisms that are not closely related. This similarity arises under the influence of the similarity of environmental conditions or a similar way of life.

Conglomerate (conglomero - I gather in a close heap) - cemented gravel. The presence of conglomerate layers and strata in the geological section indicates a stronger erosion of the older strata and the proximity of shallow water, land, or uplifts. So defines this word "Geological dictionary."

Cordayites. An extinct group of gymnosperms. They are described in detail in chapters IV and VIII.

Coating scales. Many gymnosperms (conifers, cordaites, etc.) had seeds in small bumps or on scales (seminal), which were located in the axillary of a leaf-shaped or styloid process, called covering scales or a bracelet.

Xeromorphic features. Such features of the plant that help it survive in conditions of lack of moisture are called x-rorphorphic. This is a thick cuticle, small leaves, pubescence of shoots and leaves, etc. However, sometimes the same plants have plants living in swamps. Possible reasons for this paradox are explained on pages 93-97.

Cuticle. Plants have to struggle with dryness, excess of light, diseases and pests. To do this, they have many different adaptations. One of them is the cuticle. It is a thin and elastic film of a high-polymeric wax-like substance called kutin. The cuticle is interrupted above the orifices of stomata and secretory canals. It covers the surface cells and bears their imprint, as well as their own ornaments. Having studied the structure of the cuticle, the paleobotanik can determine the plant remains much more accurately, and sometimes it is possible to find out to which group of plants the leaf, the shoot or the bud bclongs. Sometimes the cuticles of dead leaves or stems accumulate in huge quantities, forming entire layers of "leafy coal."

Lebakhia - see "Valhii."

Lepidodendron. A tree-like representative of the plumaceous Carboniferous. See the first section of chapter III.

Mangrove thickets. Dense thickets of various plants on marshy seashores. An interesting feature of mangrove plants are the respiratory roots, filled with air cavities in the roots. These thickets are formed in the mouths of rivers carrying a large amount of mud, are common only in areas with a tropical climate. In the course of geological history, the composition of mangrove plants changed. In the Carboniferous period, this was lepidodendron, calamite, some cordierite. Now these thickets are mainly composed of angiosperms.

Manoxylated trunk. If there is not enough wood in the trunk, and the core and core are the basis, it is called manoxylic. Such trunks, if they are aboveground and perennial, are reliable witnesses to the frost-free climate. Among the plants with perennial manoxyla trunks are lepidodendrons, many palms, cicadas and bennettitales (Zhang, 2001).

Maceration. Soaking of pieces of rock, plant debris, and if necessary - and parts of modern organisms in special mixtures to cause their breakdown into separate parts. With the help of maceration, the cuticle is isolated.

Megaspory. In spore plants, spores can be of the same or different size. In the latter case, out of large disputes (megaspores) female sprouts grow, and from small (microspores) - male sprouts. Of the same spores, dioecious or bisexual sprouts grow. In some plants (e.g., lepidodendrons), the germination of megaspores begins in sporangia. In gymnosperms, the megaspora is enclosed in ovum. Sporangia with microspores are called microsporangia, and with megaspores - megasporangia. Megaspores are often found in the breed in isolation; they are extracted by special methods and studied along with other plant remains.

Mesozoic era (Mesozoic). The penultimate era of the Earth's geological history. See "Geochronological scale."

Cretaceous period. The last (third) period of the Mesozoic era, which began about 130 million years ago and lasted 70 million years? In the first half of the period, among the plants, ferns, conifers, ginkgoids, cicadas, and bennettites predominate. In the middle of the period, angiosperms appeared in large numbers, at the end of the period the last dinosaurs died out on land, and in the sea - ammonites and belemnites. After that, the plant and animal world took on an almost modern appearance, but there was no human being. The period received its name from a large amount of writing chalk deposited in the seas at that time.

Mechanical fabric. To strengthen the leaves, trunks and other parts of the body of the plant is a mechanical tissue. The walls of its cells are thickened and impregnated with various substances. The cells themselves often die. 206

The microtome. A laboratory device serving for the preparation of thin and ultra-thin sections, which are embedded in preparations and are studied under an optical or electron microscope. There are microtomes, on which slices are prepared, even from spores and pollen,

Neogene period (Neogene) - see "Tertiary period."

Neotony. The ability of animals and plants to start reproducing, remaining underdeveloped in other respects, sometimes being larvae and even embryos.

Nomenclature. The whole set of names of organisms. These names are not given entirely at the discretion of the researcher. Names rules are compiled into international codes of botanical (or zoological) nomenclature, which are obligatory for all biologists. Without this, the researchers would have long ceased to understand each other. Nevertheless, amateurs violate the provisions of the code every now and then are. Nomenclature should not be confused with taxonomy.

Ordovician period (Ordovician). The second period of the Paleozoic era, which began 480 (± 20) million years ago and lasted about 60 million years. The sea basins were inhabited by a variety of invertebrate animals. Whether the land was inhabited is unclear. In the rocks of this age, unclear remains of plants are found, but it has not yet been possible to prove that they belong to the inhabitants of the land. There are no real disputes and pollen in sediments of this age. The Ordovician period is named after the ancient Ordovician tribe in England.

Paleogene period (Paleogene) - see "Tertiary period."

Paleogeography. This is the name of the discipline that studies the physical and geographical conditions of the Earth in the geological past. At the same time, the climate, the distribution of the sea and land, the relief, the character of the basins, the patterns of distribution of organisms, the features of sedimentation, etc. are considered. Paleogeography is a synthetic discipline using information obtained by various branches of science.

The Paleozoic Era (Paleozoic). The third (if we consider from our days) era of geological history of the Earth. See "Geochronological scale."

Paleozoology. The science of animals of the geological past. See "Paleontology."

Paleontology. In Greek, the "palaeos" is ancient, and "ontos" is a being. From this name, it follows that the sphere of influence of paleontology includes both the animal and plant world of the geological past. So many experts understand the scope of paleontology. In any case, all-Union paleontological society consists of individuals engaged in all fossil organisms. However, often speaking of paleontology, we have in mind only paleozoology, which of course is not fair.

Paleofloristics . It should not be confused with paleobotany. Paleofloristics is a part, and paleobotany is a whole. Paleobotany studies everything: the plants themselves, and their family ties, and their distribution in space and time. Paleofloristics studies fossil flora, the former vegetation, and not the plants themselves (Zhang, 2001).

Paleofloristic boundaries . Each flora, whether modern or fossil, can not spread all over the globe. Natural causes (climate, topography, soils, etc.) put the limit or the boundary of its distribution. As applied to the past, such a boundary is called paleofloristic.

Palynology . In addition to the prints of leaves, stems, seeds, cones and other remains visible to the naked eye, microscopic shells of spores and pollen often remain in rocks. Their study deals with the science of palynology. This science appeared at the beginning of this century, although it got its name much later. Only in the 1930s, the extraction from the rock and the study of spores and pollen were put on a wide scale and began to be applied on an equal basis with other paleontological methods for the purposes of stratigraphy. Now a rare geological department or a geological institute does without palynologists. In our country there are several hundred people.

Peristome venation . In the most ancient plants veins in the leaves were located fan (fan vein). Then there were leaves, in which there is an average vein. It leaves simple or branching lateral veins. This venation is called pinnate. Then lateral veins began to be connected by crosspieces. This is mesh venation.

The Permian period (Perm). The last period of the Paleozoic era. It began 273 (± 10) million years ago and lasted about 45 million years. During this time powerful strata of salt in the Urals, coal in the Pechora Basin and Siberia were deposited. Its name dates from the former Perm province. About plants of the Permian period is told in Chapters IV, V and VII. Together with these plants the land was inhabited by numerous reptiles and amphibians, many insects (among them beetles, cockroaches and dragonflies). In the sea, then lived many diverse animals (bivalves, corals, sea urchins and sea lilies, ammonites, etc.).

Pycnoxylated trunk . It is such a trunk in which the wood is much larger than the bark and core, that is, as in fir, oak, birch and other most common trees. Pycnoxyl trunks are not only stronger than manoxylic trunks, but can also bear severe winters if necessary.

Plain shaped . One of the types of higher spore plants. It includes the usual forest plains, which is described in the penultimate section of chapter XI, as well as giant lepidodendrons. Plain-like - a very ancient group of plants; Perhaps they were among the pioneers of land. In the Carboniferous period, they formed whole forests almost all over the world. In the Mesozoic and Cenozoic forests, they have gone to the background.

Angiosperms (flowering). First look at the meaning of the terms of ovum and gymnosperms. Angiosperms, unlike gymnosperms, are hidden in carpels, which in turn sometimes fuse, forming an ovary. The question of the origin of angiosperms is one of the darkest and most complex in paleobotany. He was even referred to the great mysteries of nature. Some ways to solve this puzzle are described on pages 160-163.

Order . What the botanists call order, zoologists call the detachment. This is a subdivision of the class.

The fern . In the subclass "Ancestral ferns" included the oldest ferns, extinct at the end of the Paleozoic. Some of the ferns are described in the third section of chapter II. These plants appeared in the Devonian period, and some of them were initially mistaken for the oldest arthropods.

Preadaptation . Adaptation means adaptation. Preadaptation (literally) is an adaptation in advance. Of course, we are not talking about the fact that organisms are far-sighted. The theory of preadaptation is based on the notion that organisms are not so much adapting to the environment as they are searching for their environment to taste. More details about this not-devoid of theory are stated in the section "Background or driving force?" Chapter XII.

Cuticle preparation . Look at the meaning of the cuticle first. This thin film, removed from the fossil plant residue, is embedded in a mixture of glycerin and gelatin, clamping it between the objective and cover slip. The resulting preparation is examined under a microscope with magnification up to 1000 times.

The Prophet . A young plant that has just grown from a seed, but already has a root, a stem and the first leaves. Not to be confused with sprouting.

Psilophytes . In Greek, "psilos" is naked, "phyton" is a plant. These are the oldest and most primitive of the higher spore plants. It is possible that they started settling the land. But there are other opinions, which are described in chapter I.

Pteridosperm . These plants, which are devoted to the section in chapter III, are sometimes called seed ferns, although it is quite possible that they had little to do with real ferns. In any case, they arose irrespective of all groups of modern ferns and are related by kinship only to prapods.

Nevertheless, the expression "seed ferns" and "pteridosperms" (in Greek "pteris" - fern, "sperm" - seed) have become firmly established.

Pollen . In seminal and some other plants microspores germinate, while still in microsporangia. The cell of the spore experiences only a few divisions, and yet it is already a real male growth. Its further development occurs in the ovum or its veils. In the end, pollen develops spermine (sperm), fertilizing the egg. Pollen has a very stable chemical shell, which has been preserved in rocks for millions of years. Next, see "Palynology."

Variety . The basic systematic unit is a species. But within the species, it is sometimes possible to identify smaller units, for example varieties.

Raznosporovye plants . These are plants that have two kinds of disputes - micro- and megasores. When there was a spasm of ambiguity, we do not know, but in any case it happened not later than the early Devonian.

Vegetation . In the middle of the last century the Swiss botanist Thurman called on his colleagues not to confuse the concepts of "flora" and "vegetation." Flora - the systematic composition of plants of any territory or geological epoch, is, in effect, a list of genera, species, etc. Vegetation is a more complex concept (Zhang, 2001). By it is meant the sum of plant communities, the general physiognomy of the vegetation cover. If we say "marsh flora," then we mean those genera and species that inhabit the swamp. Saying "marsh vegetation", we mean that at this point the vegetation cover has such a face, such a ratio of individual plants, etc.,

Replicas . When you need to study the microrelief of the object at high magnification, it's best to get a replica from it. To do this, a drop of a special solution (such as a pharmacy collodion or nail polish) is applied to the object, which dries up and turns into a thin transparent film that repeats even the finest details of the relief. The film (replica) is placed on a slide and examined in a microscope.

Semeemplating . Seed plants in megasporangia have only one megaspora. It sprouts and forms the female growth in the megasporangia, which is called a nucellus in seed plants. Above the Nucellus there are also other integuments. The inner one is called the integration. He has on the top a hole through which pollen penetrates. Above the integument there is also a cover in the form of a cup, it is a cupula (a similar shell is found in a hazelnut). All this cunning construction around the megaspora is called the ovule. Semenovels can sit directly on leaves (in pteridosperms), singly on shoots (in Buriadia, see Chapter VIII), in cones (in conifers, cicadas, etc.) or flowers (angiosperms). (Bazzaz, 2001)

Family . A systematic category consisting of genera, and in turn making orders.

Seed scales . Female cones of gymnosperms consist of a series of scales on which spermatozoa or seeds sit. These scales are called seminal. To tell how they happened, we would have to repeat Chapter VIII.

Seed ferns - see "Pteridosperm."

A seed . Semes and seed are different things. Not every semeembrant is brought to be a seed. For this, pollen must penetrate into the ovum, from which spermatozoa are formed (spermatozoa). They fertilize the egg. A fetus develops out of it. Only now the ovum has become a seed that can be sown. By the way, most Paleozoic seeds that fall into the hands of paleobotany are not rightly called seeds. It's still ovules. They never find an embryo. When they became seeds, we do not know.

Mesh venation - see "Peristome venation."

Silurian period (Silurian). The third period of the Paleozoic era, which began 422 ± 15 million years ago and lasted 30 million years. Living beings in the Silurian inhabited mainly the seas. But there were already various terrestrial plants. In the rocks of this age, there are remnants of spores. Numerous invertebrates continued to exist in the seas, but there were also ancient fish. Its name dates from the ancient tribe of the Silurians, who lived in Wales (Bazzaz, 2001).

Systematics . Spider is known for hundreds of thousands of plants and animals. It is very difficult to understand all this diversity. The task is to decompose all the organisms in shelves in the most natural and convenient way and take on the taxonomy. It is necessary to single out systematic units (species, genera, families, etc.), describe them, depict the described organisms and properly dig them. All systematic units are called taxa. The laws of taxa allocation are determined by the taxonomy discipline (not to be confused with the nomenclature). Often you can hear that taxonomy is a descriptive science and outlives its age. The answer to this thesis is contained in chapter XIII.

Systematic signs . Tick called all the features of the plant, which can be used in taxonomy. It is quite clear that if one tree has all the boughs intact, and the other has a circumcised too zealous gardener - this is not a systematic sign. With the exception of such cases it is almost impossible to say which feature is important for taxonomy, and which is not. Before we come to the conclusion that the feature is immaterial, a thoughtful researcher will measure it seven times.

Vessels - see "Tracheids."

Specialization . The division of labor and the pooling of efforts have long been adopted by the organisms. Adaptive to a narrow range of duties (that is, specialize) and individual organs (in the earthworm the entire digestive system has one gut - it and the esophagus, and the stomach, and the pancreas, the appearance of these organs - specialization), and whole organisms (a good example - Division of labor among ants). Specialization is not always beneficial. Sometimes the changing conditions of life make organs unnecessary, deprive the organisms of the usual sphere of activity. Who has gone too far in his specialization and can not restructure is for the most part doomed to extinction (Zhang, 2001).

Sporangia . The bags in which the spores are placed are called sporangia. Microsporangia filled microspores and megasporangia - megaspory. Sporangia are located singly or in groups (srusas) directly on leaves (for example, in ferns), sometimes sit on a stalk directly on stems (for example, in some plauniform or collected in cones (in horsetails, some plauniform, etc.) .The ripe sporangia crack. Sometimes a special mechanism, consisting of a group of cells with thickened walls, contributes to the disclosure of sporangia, and in some plants sporangia coalesce and form synangia.The structure of all these organs is one of the most important systematic features. ii sporangia, sori and sinangiev not always easy to understand.

Disputes . All spore plants reproduce by spores, i.e. microscopic cells formed in special receptacles (sporangia). First, a compact mass of cells is formed in sporangia, which then breaks down into small groups (usually 4 cells each). From each cell of such a group a dispute is formed. It is covered with a chemically very stable shell. After ripening and cracking sporangia, spores are sown, sprouted and sprouted. Sometimes spore germination begins in sporangia.

The stratigraphy . The life of the Earth is very rich in events. Mountains are erected and washed away, the seas come and go, forests grow on the place of deserts, rivers change their course, etc. All these events somehow affect the geological aspect: subsidence has begun - more precipitation is deposited, the upsurge began - precipitation began to be washed away. At the same time in different places of the district can accumulate quite different rocks. Not having understood the whole sequence of events, there is nothing to be taken for the search for minerals. After all, geology is not a treasure hunt for simple luck, but science! Everything has its own regularities. The stratigraphy deals with clarifying the sequence in the accumulation of rocks, comparing the thicknesses of different regions in time, drawing up and improving the geochronological scale. Every geologist should be stratigraphy, before becoming a geologist any other profile. A geologist who does not know stratigraphy is like a criminalist who is not interested in the succession of events, but cares only about the incident itself (Zhang, 2001).

Tyllites . This is what glacial deposits of past geological epochs are called. Externally, they are similar to conglomerates, but there are several signs by which it is possible to distinguish between Tillites and conglomerates. For example, boulders in the tillites carry characteristic scratches. Such glacial deposits are known in the Precambrian and Upper Paleozoic deposits.

Tracheids . A lot of terrestrial plants carry water along the trunk by tracheids. These are strongly elongated cells with thick walls adorned with spiral, ringed or ladder thickenings. The most highly developed tracheids in the thickened walls remain differently located pores. During the evolution of the tracheid, their end walls dissipated. So the vessels appeared. They are particularly common in angiosperms, but occasionally found in other plants.

Tertiary period . The term "tertiary period" has been used less and less in the geological literature since it was proposed to divide it into two separate periods - the Paleogene (Paleogene) and the Neogene (Neogene). In the old geochronological scale, the whole history was divided into four periods: primary, secondary, tertiary and quaternary. Then the first two periods turned into the Paleozoic and Mesozoic eras with their periods. The names "Tertiary" and "Quaternary" became anachronistic. Soon it became clear that the duration of the Tertiary period was quite high (it lasted about 65 million years and ended 1,5-2 million years ago), and there were many major events during it to break it into two separate periods. And they did. In the Tertiary period, the plant and animal world took a completely modern look, although man had not yet appeared. Before the beginning of the period on land the last dinosaurs disappeared, the era of reptiles finally ended. In the Tertiary period mammals dominated. In the Tertiary seas there were no more ammonites and belemnites. Among terrestrial plants angiosperms were finally established as the dominant class.

Triassic period (Triassic). The first period of the Mesozoic era. It began 230 (± 10) million years ago and lasted 45 million years. It was a time of interesting events. At the beginning of the period, different climatic zones are still clearly visible, which are greatly smoothed by the end of the period. Amazing constancy of flora on the globe - the main evidence of climate alignment. In the Triassic, the main Paleozoic groups of plants die out: the Cordaites, tree-like plaques, prairat, Kalamites, Klinolists, etc. But new groups of ferns, pteridosperms, ginkgoids, cicadas, bennetttspovyh, conifers, etc. appear in large numbers. About this, In many respects still incomprehensible change of scenery is told in chapter X. The name of the period was received in Central Europe (GDR, Germany, Austria, Switzerland), where its deposits are clearly divided into three parts: the middle one - of marine origin, the first and last - Continental. By the way, the deposits of the Permian period in the last century were called "diaz", as they are divided into two distinct parts in Europe (Lott, Manchester, & Dilcher, 1998).

Coal buds . Sometimes they are also called "coal balls, or" peat dolomites. "These are calcareous rocks in coal seams containing remains of fossil plants, surprisingly well preserved, but there are kidneys without plant remains with only marine fauna. See the first half of Chapter II In most cases, the body of the plant is covered from the outside with a cuticle or a protective film of a different type, but it is still necessary to breathe and remove excess moisture, for this purpose stomata are microscopic holes bordered by two mobile cells. A small relay has become too humid - they open and become dry - they close (the "duty" holes always remain) .The structure of the stomata and their location is the most important systematic feature.The evidence is given in the second half of Chapter II and in Chapter IV (Lott, Manchester, & Dilcher, 1998).

Phytolemma . Sometimes on the breed there is only an imprint of a leaf, a stem or other part of the plant, but more often a coal crust (phytolemma) is preserved on the print, into which the plant has turned. If this crust covers the imprint of sporangia, it can find controversy. A cuticle preparation is prepared from the phytolymph. Sometimes the phytolamps are kept surprisingly well and can be collected on exposure in herbarium folders.

Flora - see “Vegetation.”

Shaping . This is roughly the same as speciation. See “View”, as well as chapter XII.

Coniferous . One of the orders of the class gymnosperms. For coniferous, for example, spruce, fir, pine, larch. Most conifers have the appearance of needles, but conifers (modern and fossil) with wide lanceolate leaves (similar to leaves of the lily of the valley) are known. Conifers appeared in the Carboniferous period, but were especially widespread in subsequent periods.

Cicadas . One of the orders of the class gymnosperms. Cicadas have rather large feathery leaves. The old name of the cicadas is “sago palms.” Cecadian semen are sitting in cones or on mutated leaves. Reliable remains of cicadas are known from Triassic deposits, but leaves that may well belong to this order are found in Permian and even Carboniferous sediments. Cicadas dominated the tropical forests of the Mesozoic era together with the Bennettites. Leaves of both groups without preparation of cuticle preparations are often difficult to distinguish.

Quaternary (anthropogenic) **period** , or anthropogen. This period, belonging to the Cenozoic era, began 1.5-2 million years ago and has not yet ended. It was singled out thanks to two main events: the emergence of man and a major glaciation. The latter swept vast territories in both hemispheres of the planet. The glaciers of Antarctica and the Arctic are the remnants of this glaciation. The glaciation developed in several phases, and it is not clear whether it has finally decayed or even disturbed humanity (Lott, Manchester, & Dilcher, 1998). The animal and plant world of the Quaternary period, in essence, is the same as the modern, but glaciations forced animals and plants to move from place to place. On the origin of the name “Quaternary period”, see “Tertiary period.”) The Greek word “anthropogen” is based on the Greek words “anthropos” - man and “genos” - birth.

The arthropods . The type of higher spore plants, the most important feature of which are articulate shoots and whorled leaves. Sporangia are usually collected into more or less compact cones, but sometimes located directly on the stalk between whorls of leaves. Appearing in the Devonian period, arthropods (calamites, spheiofillas, fillateks, etc.) played an important role in the landscapes of the Carboniferous and Permian periods. In Mesozoic deposits, they are less common, and only one genus reached the Cenozoic era. This is a horsetail, whose history is told in the penultimate section of chapter XI. The horsetail is a typical living fossil.

Shlif . To study the microscopic structure of the rock or fossil, a section is prepared from it. A piece of rock is leveled on one side with a grinding wheel. Obtained by a flat area, the rock is glued with Canadian balsam or other resin to the glass. Then the rock starts to grind from the back side until it leaves a thin (several hundredths of a millimeter) transparent layer. This is the section. It is studied under a microscope.

Epidermis, or skin . So called the surface layer of cells protecting the leaves, seeds, stems and other organs of the plant. Cells of the epidermis, including stomatal cells, are imprinted on the cuticle. Studying the cuticle, we, in essence, study the structure of the epidermis (Lott, Manchester, & Dilcher, 1998).

Epidermal (or epidermal-cuticular) method. Study of the cuticle of fossil and modern plants for systematic purposes. This method is especially important in paleobotany. Without studying the cuticle, it is impossible to accurately determine the genus and species of the majority of fossil cordas, conifers, ginkgoids, cicadas, beepnettes and other plants. Paleobotany, neglecting this method, runs the risk of falling into the trap, framed by simulating forms. It will not only suffer (the error will open sooner or later), but also a geologist who will believe in a false definition.

The Jurassic period (Jurassic). The second period of the Mesozoic era. This period represents the whole era, it represents it in its pure form. At that time all relics of the Paleozoic forests were extinct and pioneers of the Cenozoic flora and fauna did not yet appear. The Jurassic period began about 190 million years ago and lasted about 60 million years. In the Jurassic seas lived a mass of ammonites and belemnites, quite modern in appearance fish, swam ichthyosaurs and plesiosaurs. On land, dinosaurs wandered, in the air flying pterodactyls and pioneers. To the services of herbivores there was a rich vegetation composed of ginkgo, cicadas, bennettites, ferns, conifers, etc. Judging by the plants in the Jurassic period, the climatic zonation clearly manifests itself. Jurassic deposits are widespread throughout the globe.

REFERENCES

1. Lott, T. A., Manchester, S. R., & Dilcher, D. L. (1998). A unique and complete polemoniaceous plant from the middle Eocene of Utah, USA. *Review of Palaeobotany and Palynology*, *104*(1), 39–49. https://doi.org/10.1016/S0034-6667(98)00048-7
2. Bazzaz, F. a. (2001). Plant biology in the future. *Proceedings of the National Academy of Sciences of the United States of America*, *98*(10), 5441–5445. https://doi.org/10.1073/pnas.101093298
3. Zhang, Z. (2001). [Analytical study on characteristics of medical thinking in the Sui and Tang dynasties]. *Zhonghua Yi Shi Za Zhi*, *31*(1), 21–26. Retrieved from http://www.ncbi.nlm.nih.gov/entrez/query.fcgi?cmd=Retrieve&db=PubMed&dopt=Citation&list_uids=11877106

INDEX